MW00473139

Pushing the Envelope

PUSHING THE

The Career of Fighter Ace

Maj. Gen. Marion E. Carl, USMC (Ret.),
with Barrett Tillman

NAVAL INSTITUTE PRESS
Annapolis, Maryland

ENVELOPE

and Test Pilot Marion Carl

LIBRARY OF CONGRESS CATALOGING-IN-PUBLICATION DATA

Carl, Marion E., 1915–
Pushing the envelope : the career of fighter ace and test pilot Marion Carl / Marion E. Carl, with Barrett Tillman.
p. cm.
Includes bibliographical references and index.
ISBN 1-55750-116-5
1. Carl, Marion E., 1915– . 2. Fighter pilots—United States—Biography. 3. Test pilots—United States—Biography. I. Tillman, Barrett. II. Title.
UG626.2.C374A3 1994
358.4'34'092—dc20
[B] 93-36021
CIP

Printed in the United States of America on acid-free paper ♾

9 8 7 6 5 4 3 2

First printing

Title page photo: Marion Carl set the world speed record on 25 August 1947 in this Douglas Skystreak (D-558-I) at Edwards Air Force Base. (Douglas Aircraft Co.)

For Edna, of course

Contents

Foreword

Occasionally I am asked, "What is Marion Carl like?" It is not a difficult question to answer. He is honest, direct, loyal; living testimony of the family-farm lifestyle and the Oregon earth from which he sprang. He is a primal hunter, an essential American type; he will stalk virtually anything, man or beast. His Nordic bloodline imbued him with the physical stamina and emotional stability to thrive in combat when many aviators were taxed merely to survive. That type of genetic makeup produced similar warriors named Bong, Foss, and Vejtasa.

Since 1973 I have learned a great deal about Marion. His favorite ice cream is pistachio. At age seventy-eight he still likes to drive fast. He insists that he has a poor memory (he cannot recall hearing about Pearl Harbor), yet he relates most of his combats with stop-frame clarity—perhaps because World War II was the defining experience of his life.

Like many accomplished combat aviators, Marion chose test flying as the most challenging endeavor available when peace had returned. That he was able to pursue a career on the leading edge is due in no small part to the support he received from Edna, whom he married as a nineteen-year-old Powers model in 1943. Their personalities could hardly be more different, and many friends attribute the Carls' fifty-year marriage to the conventional wisdom that opposites attract. Marion states flatly that Edna got him his first star, for she draws a crowd merely by entering a room. Yet she also possesses a wisdom to match her extroverted personality. My favorite example of Edna's priorities: "Marion, flying is all right, testing is all right, *war* is all right. But get rid of that damn motorcycle before you kill yourself."

Although best known for his fascination with exotic aerial machinery, Marion is one of the most competent and well-rounded people I have ever known, and definitely the most meticulous. He can weld steel, fell a tree, butcher an elk, wire a building. But as an aviator, those who flew with him are unanimous: Marion Carl was The Best.

Yet the man appears almost totally lacking in ego: unusual among flag officers, rare among test pilots, and nearly impossible among fighter aces.

However, he is intensely competitive—not only with others but with himself. After he acquired a replacement hip at about age seventy-three, he soon was taking long walks and climbing hills, still pushing the envelope, meeting his own standards.

He has held the world's altitude record and speed record; he has led men into battle in two wars and one clandestine operation. He retired with two stars, and when he took them off they stayed off. When he meets people for the first time, he describes himself as "retired military" rather than as a former major general of marines. He answers the phone with "This is Marion." I have never heard him refer to himself as "General Carl."

The fact that he never acquired a nickname is significant. If he were flying jets today his call sign wouldn't be "Stretch" or "Killer"—he would still be "Marion." He is not a chest-thumper. In fact, his way of expressing himself is decidedly downbeat. While transcribing fifteen hours of tapes, it occurred to me that I was listening to Pat Paulsen reading from *Yeager*. Marion Carl describes aerial combat and milking cows in exactly the same tone of voice. Therefore, if the narrative of his exceptional life seems curiously routine in spots, that is a reflection in the mirror of his psyche.

Marion has been called "the ultimate fighter pilot" and one of the three finest naval aviators of his generation. Gratifying as these accolades must be, he will tell you that they are, after all, merely opinions. His inner strength—the Scandinavian steel beneath the velvet glove—is what matters. He has nothing to prove to anyone. Least of all to himself.

But more than that, he is just the nicest man. I am glad he is also my friend.

Barrett Tillman

Pushing the Envelope

Prologue

We had known for ten days that the Japanese were coming; not only those of us in Marine Fighting Squadron 221, but the entire garrison of Midway Atoll. Navy, Marine Corps, and Army Air Force personnel were crammed onto Midway's two small islands, preparing to meet the expected enemy invasion fleet early that morning—4 June 1942.

VMF-221 had been deposited at Midway in December 1941, immediately after the fall of Wake Island. The squadron, originally intended to reinforce Wake in the face of powerful Japanese attacks, arrived only hours late. Even in those days Marine Corps aviators were carrier-qualified, and we had been preparing to launch from the USS *Saratoga* (CV-3) when word came that Wake had fallen. With nowhere else to go, we were sent to Midway, and there we sat for the next six months.

The routine at Midway had been unrelenting and monotonous, broken only occasionally by a passing enemy patrol plane or submarine. The living conditions were primitive, consisting mainly of underground dugouts. But the past several days had brought a whirlwind of change and activity. Maj. Floyd "Red" Parks had taken over upon arrival and moved up to command the squadron. He was a redheaded officer, busy coordinating air defense with our parent organization, Marine Aircraft Group 22, and with the island commander. Arrival of seven Grumman F4F-3 Wildcats had boosted our fighter strength to twenty-eight, but as of the night of 3 June one Wildcat and two Brewsters were down for repairs.

Red Parks had assigned the F4Fs to the squadron's fifth division under Capt. John F. Carey. I was a section leader in his division, a newly promoted, twenty-six-year-old captain myself. I had flown Wildcats briefly at North Island Naval Air Station, San Diego, back in August, but by the fourth of June I still had only twelve hours in type. All our F4Fs were Navy castoffs; mine sported a black number twenty-four on the fuselage and cowling. It had flown with three carrier squadrons since delivery to the Navy in December 1940.

Despite so few hours in F4Fs, I felt confident about the upcoming battle. Not so much for the overall outcome, but for myself. Having been an instructor at Pensacola, with a prior tour in a Grumman F3F squadron, I had 1,400 hours' total flight time—far more than most of my squadronmates. I knew I could fly and I knew I could shoot, so I felt prepared for combat.

However, I had secret reservations about many of -221's pilots. Some of them had only a few hundred hours' flight time and a few were unenthusiastic about flying fighters. Additionally, they were mounted on obsolescent aircraft. I felt little preference either way for the F2A or the F4F, primarily because I had some two hundred hours in Brewsters and only twelve in Grummans, but I considered myself fortunate to have drawn a Wildcat. It was a better gunnery platform than the Buffalo, and I knew that a fighter pilot is an aerial marksman before anything else.

As the sun rose I was strapped into my cockpit in readiness for the Japanese air strike we had been told to expect. Four enemy aircraft carriers were supporting the landing force of some 5,000 naval infantry—Japanese marines—and Midway's commanders felt certain the first blow would come at dawn. They were right.

At 0545 one of the PBY-5 patrol planes on its morning search reported the initial contact: "Many planes heading Midway, bearing 320, distance 150." A few minutes later the radar station on Sand Island also acquired the Japanese, reporting their altitude as 11,000 feet. The air raid alarm was sounded, and just for good measure the command post pickup was sent scurrying around, its siren wailing.

We immediately started our engines, which had been warmed up when we manned aircraft almost an hour before. There was no briefing, no coordination—just a mad scramble to get out from under whatever was inbound. Using two of Eastern Island's three runways, the Buffalos and Wildcats narrowly missed one another at the intersection. It was 0600 as Red Parks, John Carey, and the other division leaders lifted off, heading out to the northwest five minutes after the alarm had gone up.

There's an old saying that no battle plan survives contact with the enemy. That certainly applied to VMF-221, for our tactical organization broke down before the atoll was a few miles astern. The Wildcats bobbled in their climbs as each pilot cranked rapidly with his right hand—twenty-eight turns—to raise his wheels. I looked for Carey's number two man but couldn't see him. My own wingman, Lt. Clayton Canfield, was in position with me but I waved him ahead to support Carey. I continued my climb alone as tail-end Charlie.

The fast scramble had disrupted the intended squadron organization, but that mattered less than the fact that Midway's fighter squadron was airborne. We were given the vector for intercept and proceeded outbound at

14,000 feet. In all, twenty-five aircraft were straggling into a beautiful blue morning sky with almost unlimited visibility. A low-lying deck of puffy clouds partially obscured the ocean, and there was high scattered cumulus, but otherwise the weather afforded excellent prospects for early sighting and interception.

However, Carey, Canfield, and I realized we were missing our other three Wildcats. One had joined the Brewsters and two were on patrol prior to the scramble. We were alone in a hostile sky.

Twelve minutes after takeoff I heard John Carey's radio call: "Tally ho! Hawks at angels twelve." Glancing down, I spotted them—immaculate ranks of Japanese carrier-based bombers approaching Midway from 12,000 feet, barely forty miles out.

We were looking at the upper formation of Nakajima B5N bombers that had begun launching from their ships at 0430. Four aircraft carriers—all veterans of the Pearl Harbor attack—had sent 108 aircraft against Midway. One of the level bombers had aborted with engine trouble, but the other 107 came straight on in. Flying three-plane sections in V formations, the Nakajimas from His Imperial Majesty's carriers *Hiryu* and *Soryu* temporarily were vulnerable.

It was a beautiful setup. With a 2,000-foot altitude advantage our three Wildcats were well positioned for overhead gunnery runs on the bombers. No enemy fighters were evident as yet, for unaccountably the twenty-seven Mitsubishi Zeros escorting the raid were positioned astern of the bombers.

The overhead probably was the most difficult of all gunnery patterns, but it was the most effective against bombers. It gave the fighter pilot a full-deflection shot from directly above the target, depriving bomber gunners of a good angle for return fire. Few pilots completely mastered the overhead, but it was my favorite. I rolled into a 180-degree left turn, completing a half-roll to inverted.

At that moment I noticed Japanese fighters diving on me. But I was committed to my attack, knowing the Zeros couldn't hurt me yet. I jockeyed stick and rudder to line up one of the Nakajimas in my reflector gunsight, positioning the illuminated reticle to allow full deflection. My trigger finger was tensed on the stick grip as I allowed the target to grow larger in my windscreen during the 300-knot vertical plunge.

In the few seconds remaining, my total attention was devoted to the gunnery problem. I didn't know that John Carey already was in agonizing pain from bullet wounds in both legs. Nor did I know that Clayton Canfield was flying a battered aircraft that would collapse on landing. And I certainly didn't know that fourteen of my squadronmates were going to die in the next fifteen minutes.

I pressed the trigger and felt my four .50 caliber guns recoil.

CHAPTER ONE

Beginnings

I was born on a homestead near Hubbard, Oregon, on 1 November 1915. The family, including an older brother, had moved to the small farm community north of Salem in the Willamette Valley earlier that year. It was the end of a lengthy migration.

My parents, Herman and Ellen Carl, had settled on the Coquille River, in southwest Oregon, where my father had moved from Iowa. Both had been born in 1878 and possessed a lifelong commitment to hard work. Once established near Hubbard, my dad quickly became recognized as the hardest-working human in the area. From adulthood he probably never rose later than four o'clock in the morning.

My mother was equally committed. A large, extremely capable woman, she was one of eleven children and a registered nurse. Her medical training must have been useful when she delivered me that November. The family had moved onto its newly acquired land not long before, and the first priority was a barn for some forty-five dairy cattle. We lived in a tent with a wooden floor and sides with canvas over the top until more permanent housing could be provided.

Winters in the Willamette Valley are not especially harsh, but they take their toll. Rain, mud, leaden gray skies, and an oppressive monotony persist for months. This environment, coupled with the daily needs of tending the herd and milking chores, helped to breed stamina. It wasn't an easy life, yet it wasn't particularly difficult from my viewpoint. My folks expected their children to go to college, and all three of us did—perhaps the only full family of kids from our community who gained college degrees.

I admit to having been an inquisitive, accident-prone boy. At age five or six I became wedged under the barn, investigating something or other, and had to be pulled free. I tore off each ear in separate farming and playground accidents, but Hubbard's resident doctor sewed the dangling flaps back on with no lasting impairment other than a slight forward sweep. Chopping wood offered even greater latitude for disaster. I axed myself on two occasions with irreparable damage to two pair of boots. I don't know if my father was more upset about the boots or about my just being careless. I also fractured an ankle twice, and to this day it is much larger than normal. It required some fast talking on my part to pass my first flight physical.

Though intensely hardworking, my dad possessed a violent temper. It was a trait inherited by my brother Leland, eight years my senior. Naturally, this fire-and-ice combination led to conflict, and Leland—nicknamed "Jap" for his short, stocky physique—ran away from home at least twice. In 1925, when he was a high school senior, a new Model T Ford that he was driving flipped on a narrow country road. The car rolled over, pinning Leland beneath. He lived four days after the accident. I was ten at the time.

Meanwhile, the family had grown. My younger brother, Manton, and my sister, Virginia, had been born on the farm and also shared the chores. Additionally, my cousin Erma Price came to live with us; in fact, my folks had wanted to adopt her. She was a very pretty girl whose parents were separated. She stayed for a long time, so our house was always full.

We went to the same two-story building for both grade school and high school. It was a twice-daily two-and-one-half-mile walk—a dreary trek in the wintry Willamette Valley. Later Manton and I "chauffeured" Virginia on the handlebars of our bikes, which was quite a chore in wet weather and on muddy roads.

We kids were accustomed to helping care for forty-odd cows after school and all day on weekends. We did the milking by hand with the help of one hired man. Even with the extra help, milking was an all-day task, lasting from 4:00 A.M. to 6:30 P.M.

My mother was no less occupied. Aside from running the household, she was up about 5:00 every morning to wash the milk bottles. At first we sold milk locally, but the business expanded to Salem and then Portland, and her cheerful efficiency seemed to smooth over the rough spots.

Some aspects of farm life appealed to me—I didn't mind fieldwork, particularly pitching hay and handling bundles of corn—but I'll admit I never developed any affection for cows. In one summer I estimate I pitched at least 100 tons of hay, about 100 tons of corn, and 400 tons of manure that went into a spreader for distribution as fertilizer.

On my eleventh birthday I received instruction from my father on the use of a .22-caliber rifle. The family always kept two weapons on the porch, loaded and ready: the .22 and a 12-gauge Model 97 shotgun. There was a fair number of pheasant and jacksnipe on the farm, and before long I was putting meat on the table. Shooting jacksnipe improved my ability with a shotgun because those little rascals got off and moved so fast in a zigzag pattern. Throughout my life I've only felt really competent at two things—

shooting and flying—so I guess it was natural for me to become a fighter pilot.

At age twelve I "officially" learned how to drive. It wasn't uncommon in those days since most of the necessary driving was limited to the farm. However, the local milkman had allowed me to drive his truck on occasion three years before. I got my driver's license at sixteen, and not only drove the car and tractor but also assumed responsibility for their maintenance. Manton, eighteen months younger than I, showed more interest in farming and would study agriculture at college, while I leaned toward engineering. As an avid tinkerer and mechanic, I was already on the path that would lead to aviation.

But there were setbacks along the way—serious setbacks. In February 1933 my father died at age fifty-five following a double mastoid operation. It hit us hard, but my mother took up the slack and continued running the family, perhaps even more efficiently than before. We knew our mother was quite resilient, a very strong person both mentally and physically. Despite Dad's death, she insisted that I pursue my plan of attending Oregon State College in Corvallis the following year.

At age seventeen I was pretty well prepared for adulthood. If I had a major fault, it was a happy-go-lucky attitude coupled with a love of speed. Part of this characteristic may have been inherited, since my mother had a pretty heavy foot herself. One wintry day, driving my mother, I intended to make the first turnoff into Hubbard when our Chevy began to skid on the ice. I instantly corrected, downshifted, and made the second intersection without difficulty.

However, I hadn't been checking my tail. A state trooper close behind me also tried to make the second turn but lost control and spun 360 degrees in the middle of the road. He proceeded into town via the third exit and tracked me down at the hardware store.

Confronted with a judge about to impose a $25 fine for reckless driving, I counterattacked. "Look," I insisted, "just because the cop couldn't make the intersection he gave me a ticket for reckless driving instead of speeding. He's the one who lost control of his car!" In retrospect, that may not have been the most diplomatic approach.

About this time the flying bug bit, and it bit hard. I always had been interested in airplanes, as the farm was on the airmail route from Portland to San Francisco. Occasionally I saw Army airplanes from Fort Lewis near Seattle, and to a youngster enamored of things mechanical, flying held a powerful attraction.

With my interest in mechanics, I opted for engineering at Oregon State. I was especially interested in aeronautical engineering, but the school had no degree in that specialty. Therefore, I became one of five students taking the aero option to the engineering curriculum. Our instructor, Ben Ruffner, had formerly been an aeronautical engineer with Boeing Aircraft in Seattle. Eventually the other four students went to work for Boeing, but I still intended to fly.

However, there were roadblocks in the way. I was a good student in high school, but in college I was in pretty fast company. This situation was compounded by the fact that I didn't like to study; I much preferred sports. I had played baseball in grade school and, at more than six feet, had been a starter on the Hubbard High basketball team. I also enjoyed boxing and wrestling and won all three of my "smoker" bouts by technical knockouts. But while these combative sports may have been just the ticket for an aspiring fighter pilot, they interfered with the tedious chore of studying. Additionally, I was on a tight budget and athletics ate into my part-time work schedule, so I dropped organized sports and worked as a waiter in my dorm for thirty-five cents per hour.

Oregon State was a land-grant college, which meant that almost every male student was obliged to participate in the Reserve Officer's Training Corps program. The Army ROTC offered infantry, artillery, and engineering. I had no choice, since engineering students were assigned to the Corps of Engineers. My brother, studying agriculture, chose the artillery branch.

After two years of college I had reached the burnout stage. I was tired of the constant studying—especially in my engineering courses—and decided to help at the farm for a year. I made two trips to California during this time, and the satisfaction I found in self-reliance stayed with me the remainder of my life. On the first trip I jumped a freight train out of Salem and showed up on my aunt's doorstep in Alameda, so sooty and dirty that she didn't recognize me at first. The second time I hitchhiked as far as Santa Ana, where I picked citrus in Orange County's groves.

Upon returning to Corvallis in 1937 at age twenty-one, I felt ready to complete college and get on with an aviation career. I had saved enough money to buy a Model A Ford roadster for $175. My foot was always either on the floor or on the brake. On one occasion I was giving a lift to another student when I heard about a fire on campus, and we beat most of the spectators to the scene. My passenger became a naval aviator and senior United Airlines captain, but years later he admitted he preferred an airplane to riding with me!

Money was tight, so I set up a cooperative with five other fellows, one of whose mothers was known by all of us. She served as cook and housekeeper during my junior year, with monthly room and board amounting to $22.50

per student. With a better handle on academics, I again went out for competitive sports: boxing, wrestling, jiujitsu, and fencing. I also developed into a strong swimmer, logging one mile three times a week.

Meanwhile, I had scraped together enough money to take flying lessons. A local instructor, Dick Linden, had an airstrip along Route 99 East, south of Corvallis. The cost was not exorbitant for those days—we agreed to twenty hours flying for about $100. Flying with Linden in a J-2 Cub, I soloed after two hours and thirty-five minutes of dual instruction. I'm still proud of that record; I seemed to feel comfortable in the air right from the start.

Linden had other aircraft that provided a little diversity. The OX-5–powered Alexander Eaglerock was a trim two-seat biplane from the barnstorming days. I enjoyed flying it, but between the Cub, the Eaglerock, and one hop in a Rearwin I depleted my cash reserve quickly. Flying low and slow over the lush greenness of the Willamette Valley and gazing down upon everyday objects from an overhead perspective was enchanting. I resolved to return to aviation, and a stint in the armed forces offered the best prospect.

Upon graduation in 1938, I was commissioned a second lieutenant in the Army Reserve as a member of the Corps of Engineers, but that was a mere formality as far as I was concerned. I had never lost my ambition to become a military pilot, and not even my crushed ankle from childhood kept me from passing the Army flight physical. Or, for that matter, the Navy flight physical. I was playing the percentages.

I was shut out of the Army flight program on a technicality—the quarterly quota at Ft. Lewis, Washington, was filled. But with a waiver for my ankle I was accepted as a prospective naval aviation cadet. Then fate intervened in an immaculate dress uniform—boots, breeches, and Sam Browne belt.

Captain Stedman was a sharp-looking Marine Reserve officer. Under a little-known act passed by Congress years before, 10 percent of Army ROTC graduates could accept Marine Corps commissions. Although I was inclined toward the Navy at first, Captain Stedman soon changed my mind. The Marine Corps was accepting only college graduates for flight training, while the Navy would accept those with just two years of college. Additionally, by selecting the Marine Corps I could get to Pensacola one month sooner than the rest of my elimination training class.

That settled it as far as I was concerned. However, I first had to deal with the bureaucratic impossibility of being listed as a member of all three armed forces. Was I an Army lieutenant, a Navy seaman, or a marine? I expressed my intent, and became a private first class in the U.S. Marine Corps.

Elimination training was held at naval air stations around the country to pare down the number of student aviators for the Navy's main flight school at Pensacola. Applicants from the Pacific Northwest went to NAS Sand Point in Seattle, where I arrived in August 1938. During my first flight with Captain Stedman, he could tell as soon as I took the controls and made a few turns that I had previous experience. He asked me if I had done any private flying, and I told him about my twenty hours as a civilian. I was the first one in my elimination class to solo. Three of the eight in my class failed; it was typical of the era when about 40 percent of each class washed out.

While in Seattle I wrecked my Model A. Or, more precisely, another driver did the job by hitting my Ford broadside. I never got compensation, but I still managed to buy a 1936 Ford V-8. I promptly got under the hood and milled the heads down to increase the compression ratio so I had quite a bit of acceleration if not a lot of top speed. I observed my twenty-third birthday 1 November, then loaded up my new car and headed for Pensacola, arriving in time for Thanksgiving 1938.

It was appropriate timing. I was full of gratitude at the recent turn of events and knew that I wanted to spend the rest of my life flying airplanes. I knew that after three years in the Marine Corps I would probably have a choice between civilian flying and continued military service.

CHAPTER TWO

Pensacola

I wasted little time making my presence known to the citizens—or at least the traffic cops—of Pensacola. Eight days after my arrival I received an off-base speeding ticket. Things got worse six days later when I lost my appointment as a wing leader when some cadets in the rear of my formation ambled along, out of step and chattering among themselves. The entire cadet wing received ten demerits each and was required to spend an extra tour of duty. Other people might have been prompted to have second thoughts about their prospective career as a result of such a setback, but I knew for certain that I could "hack the program."

The Pensacola syllabus was being streamlined for marines in 1939 as greater possibility of war increasingly became apparent. That suited me just fine. Previously, every cadet had received the full training program, passing through five squadrons for a thorough grounding in every variety of Navy aircraft. Students began on floatplanes, progressed through two types of landplanes, learned to handle twin-engine seaplanes, and finally graduated to tactical aircraft. However, because the Marine Corps had no patrol planes my colleagues and I skipped part of Squadron Four. But it took a while to get that far.

My first flight at Pensacola was in an N3N-1 floatplane in Squadron One with a Lieutenant Lindsey as instructor. That was 12 December. I soloed 5 January 1939, again first in the class, and got thrown in the bay as a result.

Things continued smoothly for the next few months. I progressed to Squadron Two, flying N3N-1 landplanes and occasionally the lighter, more responsive, Stearman NS-1. However, on a solo hop 3 May I was tempted by the prospect of a strictly forbidden dogfight with another student. We both succumbed to the impulse and were having a grand time when an instructor happened upon us. He saw my tail number, but the other culprit made a clean getaway. Upon landing, I was grounded and restricted to base; there was a chance I would be expelled from flight training.

For more than a week I was allowed to ponder the error of my ways. Then, on 12 May I got the word: fifty demerits and fifty hours marching "the area" with an '03 Springfield rifle on my shoulder. The restriction to base remained in effect, but it didn't stop me from leaving the station to honor a date. Two days later I was returned to flight status, and by the first of July I had marched off the last of my punishment tour and the base restriction was lifted.

That was the good news. The bad news came three weeks later when, as leader of a nine-plane formation, I watched two of my pilots collide. Both damaged aircraft landed safely, minus some wing parts, and all nine of us were grounded pending investigation. The inquiry took a week, and at the end of that time I was interviewed by the station commandant, Capt. Aubrey Fitch. In 1939 captains spoke only to admirals and admirals spoke only to God, so I was way out of my depth. But the result was a reprimand and return to flying. There was another minor setback when my car was impounded for a month after I was caught speeding on base again, but by then I'd grown to expect such inconveniences.

Up to this time my class had been flying obsolete fleet-type aircraft: Vought O3U observation planes, SU scouting aircraft, and the trim, semi-operational SBU dive-bomber. Although the flying boat curriculum in Squadron Four was shortened for marines, I ended up with another period of enforced grounding anyway. However, there was no disciplinary violation this time. Rather, a muscular imbalance of the eyes—exaforia—mysteriously cropped up. It proved temporary and I eagerly moved on to Squadron Five.

November 1939 was the highlight of my time at Pensacola. After a full year in training, our class finally was ready for current fleet aircraft. We flew the Douglas TBD-1, which two years earlier had become the Navy's first monoplane in squadron service. But the real thrill—the long-awaited reward for tolerating naval discipline—was flying Boeing's elegant little F4B-4. A fighter!

I flew the F4B almost fifty-two hours that month, and I savored every minute. The cramped cockpit seemed to fit my six-foot-two-inch frame like a glove. And that is how most pilots recall the F4B; they wore it as much as flew it. With exceptional power-to-weight ratio and a semi-indestructible airframe, it inspired student aviators to probe the limits of its performance—and theirs.

The unbridled joy of flying such an aircraft clinched my decision. I was absolutely determined to join a fighter squadron, and while still at Pensacola I began working toward that goal. There were others with similar ideas. For example, one month ahead of me was a big, good-looking guy named Bob Neale, and one class behind me was an even bigger fellow from Texas named David Lee Hill. They became the top guns of the American Volunteer Group (AVG) when General Chennault's "Flying Tigers" entered combat in 1941-42. Both those fellows were not only top-level individuals but top-level fighter pilots as well. It's not generally realized that about 60 percent of the AVG pilots were Navy trained.

I also made other friends, including some of the relatively few marines in flight training at that time. One in particular, a class behind me, was Art Adams. For his entire career Art was one number junior to me, the long-term effect of an accident of timing in entering Pensacola. That one-number difference followed him every day of his career, including the day he pinned on his second star three decades later!

In mid-November I received a letter from 2d Lt. Carl Longley, who had been four classes ahead of me. Longley was stationed at Quantico, Virginia, and noted a vacancy in Marine Fighting Squadron 1. He suggested I try to nail down the seat, and that letter made all the difference in how I started my career. Indirectly it led to my decision to make a career of the Marine Corps.

Meanwhile, there still was work to do. Late that month I qualified in aerial gunnery aboard the F4B—only three of us in Class 120 did so—and graduated the first of December. Freshly decked out in my first Marine Corps uniform, with second lieutenant's bars and—best of all, Wings of Gold, I immediately set course north to Quantico. No graduation leave, no detours; just straight to VMF-1 and a chance to fill the Marine Corps's only fighter pilot vacancy.

Regulations allowed five days' travel time, but I didn't take even that much. During the drive I had time to ponder the events of the previous twelve months. With two years of aeronautical engineering at Oregon State and my civilian flying experience, I had been overqualified the day I drove through the main gate at Pensacola. I analyzed my year as a cadet and decided I had had too much time on my hands, since both ground school and flying came fairly easily. I had lacked enough activities to keep busy and, therefore, out of trouble. I decided it wouldn't happen again. With 320 hours in my military logbook and a string of top grades and "up" checks, I was off to an excellent start. Now I just needed to get assigned to Marine Fighting 1.

The prospects looked dim. I arrived at Quantico as a newly minted lieutenant on 3 December and reported to the air group adjutant, Capt. "Fish" Salmon. I laid my cards on the table: I had given up my thirty-day leave and forfeited half my allotted travel time for a shot at VMF-1. He seemed more impressed by my engineering degree than by my flight grades and said, "You'll fit right into the maintenance squadron in the airbase group."

I was crestfallen. I'd just as soon have gone to a rifle platoon. It took me a while to realize that the adjutant possessed a sadistic sense of humor that was well camouflaged. Suffice it to say, that same day I joined Marine Fighting 1.

The squadron was loaded with talent. I learned the skipper was Maj. T. J. Walker, the executive officer was Capt. Ed Pugh, and the flight officer, in charge of operations, was Capt. Joe Bauer. Other pilots included Marion "Mac" Magruder, Joe Renner, and John Condon—all future standouts in World War II.

However, there were the inevitable personal and professional jealousies that dog any competitive organization, since a fighter squadron—especially a Marine Corps fighter squadron—is nothing if not competitive. I checked out in the F3F-3 on 12 December and immediately took to it—small, agile, with excess power and that radial engine sound that told everyone in earshot: this is a fighter airplane.

But the F3F could be a handful, especially on landing, and shortly after arrival I dragged a wing on touchdown at Quantico's Brown Field. I had followed too closely behind another aircraft and the slipstream got me. The damage was negligible—it was repaired in a couple of hours—but some of the old hands exchanged I-told-you-so looks. One of those was "Indian Joe" Bauer.

Fighting 1 had had some earlier problems with reservists and second lieutenants. One such, who had moved on, was a former University of Washington aero engineering graduate named Boyington. The situation was complicated by a Navy-wide change that eliminated the customary three-year requirement to serve as a midshipman. Largely a budgetary measure, it had permitted graduate aviators to fill squadron billets without being commissioned until their "purgatory" expired. Those who had come up through this system tended to resent newcomers who arrived as commissioned officers.

Other problems did nothing to alleviate matters. Some of the reservists were permitted to file applications for regular commissions about this time, drawing further resentment. Then a pilot named Caldwell turned up. He had pinned on his gold bars and his wings a month before I left Pensacola, believing he was to fly the F3F marked l-MF-18 after a month's leave. Upon arrival he learned he had been scooped and had to settle for duty in another squadron. My ambition and foresight had paid off, but I wasn't winning new friends.

Accommodations were insufficient aboard the station for all the bachelor officers, so I was billeted offbase. Driving to work the second day of 1940, a patch of frost on my windshield caught the early sun's glare and I lost all forward vision. Easing too far to the right, I wiped off that side of my Ford V-8 on the guardrail.

As if that weren't enough, a couple of days later VMF-1 lost an aircraft. A pilot named Doswell had a girlfriend who lived in the rural Virginia hills, and he dropped down to say hello. While making a low pass he hit the ground and completely wrecked the F3F, although he walked away with only slight injuries.

My luck improved at this point. As the most junior pilot in the squadron, normally I would not have made the annual gunnery trip to the Caribbean. Now, however, I filled the available seat and embarked on maneuvers, leaving Quantico the middle of January for St. Thomas via Fort Bragg, Jacksonville, Miami, and Guantánamo Bay.

For an Oregon farmboy, the Caribbean was an exotic environment. I relished the gunnery hops, but not the chores that went with them. Assigned to 1st Lt. Joe Renner's ordnance section, I drew the task of supervising the boresighting of the F3Fs' guns at St. Thomas, Virgin Islands. It was a time-consuming, tedious job. Each aircraft had to be taken to the firing point, the tail raised to level-flight position, and the guns tested for point of aim with the engine running since they fired through the propeller arc.

I watched more than supervised, as it was obvious that the gunnery sergeant knew more about the process than I did. In the hot, humid atmosphere I became bored and lay down beneath the F3F to allow the NCO to complete his work unhindered by a fresh-caught aviator. Actually, it was a time-honored means of dealing with an unfamiliar situation. Second lieutenants have been getting by for centuries by saying, "Carry on, sergeant."

Unlike most people, I'm fortunate in that I can sleep anywhere, anytime, under almost any conditions. So, despite the engine noise and occasional gunfire, I drifted off.

Next thing I knew I was being rudely awakened. There, looming above me, was Joe Renner. First Lieutenant Renner, in fact, speechless with rage. And it's not often that you found Joe Renner without a mouthful of words, especially when he was angry. He never forgot that incident—neither did I!

But there were compensations. The flying weather was fine, and the crystalline beauty of the Caribbean, with its varying hues of blue, and the cumulus-piled sky, were enchanting. The raw thrill of flying a high-performance airplane and shooting at towed sleeves made all the work and turmoil more than worthwhile.

Then it was back to wintry Virginia. To regain mobility, I went shopping for a new car. I looked up an acquaintance in Fredericksburg, Virginia, named George Purvis, who sold me a Buick Special to replace my late lamented Ford. My reasoning was that even with the V-8 I hadn't managed to avoid three speeding tickets in a year and a half. I expected the Buick's better speed to allow a margin of superiority over the opposition.

While testing this theory at 92 mph on Route One, I was distressed to note a state trooper closing on me—in a Ford. Once the formalities had been conducted alongside the road, I asked the officer how he had overhauled a Buick Special in a Ford. He unzipped a grin and said that a friend in Fredericksburg had installed a Mercury engine, and that Mr. Purvis guaranteed a top speed in excess of 100 mph. That education cost me another $25.

Fredericksburg, the scene of the 1862 Civil War battle, drew a lot of attention from VMF-1. Most of the lieutenants spent their evenings there, as the primary attraction was the Mary Washington College for Girls. While I enjoyed socializing, I had no more intention of finding a prospective wife than I did of transferring to the infantry. Flying continued to absorb me, to the point of teaching my "friend" George Purvis to fly (after a fashion) in civilian aircraft. Having applied for and received a commercial license, I enjoyed the variety of private instruction in such aircraft as the Kinner Sport, a racy-looking open-cockpit monoplane.

But all too soon the Fighting 1 dream came to an end. In late May 1940 I was notified I was being sent back to Pensacola—as an instructor. In my six months with VMF-1 I had logged some 130 hours in the delightful little F3Fs. Now, reporting to Squadron 1-A at Correy Field, I contemplated teaching cadets takeoffs and landings in the stable, sedate N3Ns. However, I did recognize the potential advantage of an instructor's tour. The old saying holds true in aviation as much as anywhere else: the best way truly to learn a subject is to teach it.

Certainly there was plenty to teach. I most enjoyed aerobatics, and later considered refinement of my "stunt-flying" skills the greatest benefit of my fourteen-month tour. Though I didn't recall the episode, years later Joe Foss told me about a ride he begged in the rear seat of a Stearman. He was an eager cadet at the time, and the way he tells it, we spent the entire ninety-minute flight upside-down or rolling through the upright position on the way to inverted. Upon landing, we climbed out and Joe says he swallowed whatever was in his mouth, managing to say something that sounded like, "Thank you, Lieutenant Carl." "Then," explained Joe, "as soon as old Marion lugged his chute back toward the line shack, I dropped on all fours and upchucked the soles of my shoes."

Other types of precision flying also appealed to me: small-field procedures and landings to a circle. The diversity of students also was noteworthy in 1940. Not only did it include the usual cadets, but also Royal Navy candidates and U.S. Navy or Marine Corps commissioned officers who had qualified for flight training. At one point I had eight students and logged eighty hours in one month. But after the first six months or so the novelty wore off and I, like many instructors, began looking for other diversions.

Inevitably those diversions assumed the shape of a busman's holiday. Small formations of instructors took turns leading groups of Stearmans or N3Ns, alternating the lead. On one memorable occasion a former VMF-1 squadronmate (the one who crashed his F3F at his girlfriend's house) pulled into too tight a loop. The leader stalled at the top and spun out, prompting similar reactions among the other wingman and myself. I still have a vivid image of three trainers dropping into inverted spins, and have often wondered if any witnesses on the ground thought the maneuver was intentional!

Pensacola's social life was considerable in 1940–41. Most of the instructors were of the same age group and quite a few were bachelors. But I was definitely uninterested in matrimony and made a policy of not dating the same girl for too long. However, despite the pleasant atmosphere and opportunities for flying, I wanted out after about six months. Much later I hitched a ride on a flight to Washington, D.C., with one goal in mind—transferring back to fighters.

The officer detailer in the aviation branch was a Major Majors, who must have been a frustrated fighter pilot. He listened sympathetically to my tale of woe; I had been instructing for a year and was likely to be left in that position for another twelve months or more. Without committing himself, Majors said he would look into the matter and let me know the results.

So I returned to my students and biplane trainers, dutifully teaching a new crop. But there was one bit of good news along the way—I was selected for a regular commission, and this was a sure sign the Marine Corps wanted to keep me.

Meanwhile, my mother and sister arrived at Pensacola on 18 June 1941 for a visit. I took some time off in order to show them around the air station and the surrounding area. They headed home on 23 June, and I caught up with them later that month in Bakersfield, California, and rode northward in Ginger's car, bound for Hubbard.

Near Klamath Falls in south-central Oregon, Ginger failed to make a curve and the car rolled at least twice. At the time I was asleep in the back with a carton of peaches beside me. When it was all over I was on the floor with the peaches on top of me. My mother and I were merely shaken up, but Ginger suffered a painful broken elbow. The car remained functional, and I drove the rest of the way home without drawing the unwelcome attention of the Oregon State Police.

Back in Pensacola at the end of July, I learned that Major Majors was as good as his word. I was assigned to VMF-221 with orders to report two weeks later in San Diego. The world looked bright again.

When I joined VMF-1 at Quantico in 1939, I believe the Marine Corps possessed only seven tactical squadrons: two bombing, two fighting, and three scouting. Now that had changed, and the corps boasted four fighter squadrons alone—two on each coast—plus expanded scout-bomber and support units.

But some things remained constant. When I reported to North Island Naval Air Station, I found Major Manley as CO of -221 and Maj. Verne McCaul as the exec. The operations officer was none other than Joe Bauer, one of the unimpressed seniors from VMF-1. But I resolved to remedy the situation. I knew that part of the resentment against me at Quantico stemmed from my status as a reserve second lieutenant. Well, now that had changed. I also acknowledged that my original attitude probably had not helped, as I attached far more importance to flying than to the other things expected of junior officers. In short, I was learning.

Things turned around at North Island, very much in my favor. I checked out in the new Grumman F4F-3, but the Navy recalled its badly needed Wildcats almost immediately, and Brewster F2A-3s were issued instead. The Buffalo became the unlikely mount upon which I established a warm relationship with "Indian Joe" Bauer.

As with any new kid on the block, a tussle was inevitable with the reigning top dog. In VMF-221 that was Bauer. We squared off well inland, east of San Diego. I was strapped in tight because I knew I was about to lock horns with the finest fighter pilot in the Marine Corps. We went at one another man-to-man in a free-for-all that would establish the squadron's new pecking order. I don't remember individual maneuvers from that fight, but I distinctly recall making a half-snap roll to recover from an inverted spin below the crest of a ridgeline, with neither of us having gained an advantage. From then on we regarded one another with mutual respect. I came to admire Joe Bauer as perhaps the finest pilot and officer I ever knew.

From August into December 1941, VMF-221 was an uncommonly pleasant assignment. The pilots esteemed the skipper and exec as real gentlemen, and Bauer took pains to bring the newer fliers up to speed in dogfighting. It had far more to do with survival than with mere proficiency or aviator ego. Joe was trying to teach us how to exist in the air, and he did a good job.

Major Manley soon was transferred out of the squadron and Verne McCaul "fleeted up" to command the unit. Popular as he was with the welfare of his troops always in mind, McCaul just wasn't much of a fighter pilot. Almost any second lieutenant could beat him one-on-one. More than once the CO returned from a fight, threw his gloves on the deck and exclaimed, "If there's only two ways to turn in a dogfight, I can always pick the wrong way!" Later I learned that most pilots who did not enjoy fighters could perform well in other aircraft—usually multi-engine types.

The squadron training program continued unabated, including section tactics, instrument practice under the hood in an SNJ trainer, and field carrier landing practice. Periodically we went to sea on board the carrier USS *Saratoga* (CV-3) to watch Navy pilots making launches and landings, as VMF-221 was scheduled to depart for Hawaii on 8 December.

Sure enough, the squadron completed training as planned and stowed aircraft, equipment, and personnel on board by the evening of the seventh. I'm slightly embarrassed to admit it, but I have no recollection of exactly where or when I heard of the Japanese attack on Pearl Harbor. I may be the only member of my generation who claims that dubious distinction, but the episode is a blank. If I have any impression, it concerns the general difference in attitude between married men and bachelors. Certainly there were exceptions—Joe Bauer was a husband and father, and he was as aggressive as they come—but I think that overall the unmarried pilots were more eager to get involved in the war, figuring that we now had a chance to do what we had been trained to do.

We boarded the *Saratoga* and she slipped away from North Island the morning of the eighth, headed west, to war.

CHAPTER THREE

Midway

The *Saratoga* sailed with eighty planes of her own air group and VMF-221's dozen F2A-3s. Stashed among our squadron gear was my Cushman Eagle scooter. I had sold my '39 Buick Special for $575 and purchased the scooter, which would do 40 to 45 mph if there was no headwind. I was probably the only Marine Corps fighter pilot with his own personal transportation in the ever-widening war zone. I crated the Cushman and it went with the squadron gear to Midway.

The big carrier put in to Pearl Harbor the morning of 15 December, and evidence of the five-week-old Japanese attack was still apparent. Sunken, shattered ships, fuel oil on the water, and damaged or destroyed buildings all attested to the fact that America was at war.

And the war news was uniformly bad. Wake Island was garrisoned by a Marine Corps defense battalion and Maj. Paul Putnam's VMF-211. Wake's defenders had held off the Japanese so far, but it was a losing battle. The *Saratoga* task force, under Rear Adm. Frank Jack Fletcher, was dispatched to relieve Wake, and the marines in "Sara" were anxious to get there. Aside from a professional interest in plying our trade, there was a personal reason as well. Nearly every pilot in -221 had former classmates, friends, or acquaintances in -211. I often thought of Lt. John Kinney, whom I had known at Pensacola. He was one of Putnam's pilots, and much later I learned that John had worked small miracles in keeping their few remaining Wildcats operational.

The *Saratoga*, her tanker, and escorts were to haul within range of Wake on 22 December. But despite originally high spirits in VMF-221, events dragged squadron morale as low as the descending clouds. The need to zigzag to avoid possible submarines, plus the slow speed of the accompanying oiler, reduced the thirty-knot carrier and cruisers to a twelve- to sixteen-knot rate of advance. Much criticism was leveled at Rear Admiral Fletcher for his cautiousness over fuel state in his escorts, and all chance of relieving Wake was lost. The garrison surrendered the morning of the twenty-second (twenty-third Wake time), and the relief force reversed course a few hours later. The *Saratoga* had fallen 425 miles short of her objective.

The only place left for VMF-221 was Midway, and we launched from "Sara" on Christmas Day. An immediate schedule of morning and evening patrols was established, with a four-plane division aloft two hours at a time. Those pilots not flying were on call from dawn to dark.

So began the drab, monotonous existence we would endure for the next six months. Living conditions were dank and unpleasant, consisting of dugouts whose floors were barely above the water table. When the wind whipped tides higher than normal, water would seep into the dugouts—sometimes to a depth of twelve inches. We pilots spent most of our days in the ready tent, trying to keep ourselves occupied between patrols.

It was no easy task. I had a Zenith multiband shortwave-longwave radio, perhaps the only one on the island. Between the radio and the scooter I had more diversions than most, but my jury-rigged radio antenna kept falling because of birdstrikes. Midway, it turned out, was a nesting ground for migratory maritime birds, the most numerous being the Laysan albatross. We called them gooneybirds. Their comical appearance and ungainly antics on the ground made them a source of fascination for aviators, and we seldom tired of watching the young birds with their fuzzy-looking, downy brown feathers.

Still, not even the gooneys could fill so many idle hours. I noticed that the city-bred pilots seemed to cope poorly with Midway's enforced idleness, while those from agricultural backgrounds fared somewhat better. But with unreliable electricity in the dugouts and a scarcity of reading material, some nerves became frazzled.

To alleviate the boredom, I tried to find things to keep interested in. Naturally, my biggest fascination was flying. On one patrol I decided to slow-roll around the perimeter of the coral reef. Since wingmen were supposed to follow their leaders, my number two had no choice but to duplicate my maneuvers. By the time we landed he was groggy and mad; he described me as a "slow-rolling so-and-so." It was a reaction I had noted previously. Not everyone assigned to fighters was enthusiastic about aerobatics, and some were uncomfortable in inverted flight. I couldn't help but wonder how such pilots would fare in combat.

Reminders of the war came Midway's direction at odd intervals. Japanese submarines lobbed a few shells at us three times during January and February, causing little damage. On the last occasion, 10 February, two of our pilots were airborne on dusk patrol when the shooting started. They attacked and dropped their light bombs close aboard the sub, which promptly pulled the plug and dived.

Better results were obtained a month later. On the morning of 10 March, Midway radar detected an unidentified aircraft forty miles west and vectored out a dozen Brewsters. Capt. Jim Neefus's division made contact with a Kawanishi 97 flying boat, probably from Wake, and all four marines made gunnery passes. As the big Kawanishi dived toward cloud cover at 3,000 feet, Neefus connected and smoked one engine. However, the Japanese were shooting as well. They put seven holes in the F2A flown by Gunner Robert Dickey, and one round through his shoulder. However, Neefus bent his Buffalo around for a second pass and sent the flying boat tumbling into the water. VMF-221 celebrated with a bottle of bourbon donated by the air group skipper, Lieutenant Colonel Wallace.

Late that month we received five new pilots, all second lieutenants. I took an immediate liking to one of them—Roy Corry, from Santa Ana, California—whom I would later fly with. At this time we were flush with second lieutenants, though we had lost some experienced pilots—including Joe Bauer—to new units. But the news of promotions on 12 February had been especially pleasant. I was one of the new "first louies," but there was a hitch. No silver rank insignia were available on the atoll, so our gold second lieutenants' bars were coated with solder. Three months later I was sworn in as a captain. Events were accelerating.

So was the pace of squadron life. VMF-221 turned over two skippers in April and May, with Jim Neefus relieving Verne McCaul, who moved up to group operations. But Neefus was recalled to Hawaii in May and was replaced by Maj. Floyd "Red" Parks, our fourth CO in seven months. Of those four, Neefus was the best fighter pilot and McCaul the worst, though Verne was a very good leader and well liked.

The new skipper was unknown to most of us, but the entire garrison quickly learned of his presence. After cockpit alert the next day, Parks warmed up his F2A and taxied out to the runway. Without filing a flight plan, he took off and logged a short hop around the atoll. Both islands immediately went to red alert, suspecting the worst.

In the squadron command post the telephone jangled off the hook. As duty officer that morning, I picked up the receiver and heard a distressed Verne McCaul on the other end, demanding to know what was happening. MAG-22 operations showed no planned flight; what was doing? I told him that the CO had taken off without notice, reason unknown. With that McCaul requested Major Parks's presence immediately upon landing. Parks quickly got the word. He told McCaul that he didn't believe in warming up an airplane without flying it. But that was the last unscheduled takeoff from Midway for a long, long time.

During the five weeks before the battle, almost every day brought new developments. Admiral Nimitz and his Pacific Fleet staff visited on 1 May and set more rumors astir. Then on the twenty-fourth all personnel were ordered to turn in or destroy their diaries for security reasons—an edict I ignored.

First warning of impending attack came 25 May, and VMF-221 began maintaining four planes on alert during daylight. It was a tiresome, boring business lasting from before dawn until after sunset. The Brewsters were reinforced two days later with the arrival of seven F4F-3 Wildcats assigned to Capt. John Carey's division. As a section leader in the flight, I traded in my Brewster for a Grumman, though there was precious little time to fly the new fighters. I don't think we ever learned why the fifth division received the Grummans, but Carey, McCarthy, and I had checked out in F4Fs at North Island, while few if any of the squadron's other pilots had ever flown Wildcats.

More help arrived over the next week: eighteen Douglas SBD-2 dive bombers for VMSB-241, a dozen PBY-5As of Patrol Squadron 44, plus Army B-17s and B-26s. Sand and Eastern Islands were crammed with aircraft, almost displacing the gooneybirds as Midway's aerial denizens.

On 1 June came word of a Japanese air attack against Dutch Harbor, Alaska. Clearly things were heating up, and next day Midway's last reinforcements touched down: six Grumman TBF-1s from Torpedo Squadron 8. The big Avengers were too late to sail on the USS *Hornet* (CV-8) with the rest of the squadron, so they made a last-minute flight from Pearl Harbor to Midway. Somehow the Navy aircrews settled into the scant room left amid parked aircraft and freshly built revetments, then joined the waiting game.

On 3 June the Japanese task force was reported to be closing in. Midway braced itself. Status: invasion imminent.

Diving vertically past the Nakajima level bombers, I was far too busy to check the score. I neither saw nor claimed any damage to my intended victim, as I was more concerned with the Japanese fighters that suddenly appeared above me as I had begun my run.

I recognized the enemy aircraft as Mitsubishi Zeros. Though we didn't know much about the Imperial Navy's premier fighter, a few things were evident. The Zero was at least as fast as the Wildcat, it climbed far better, and it had superior maneuverability. So I remained in my headlong plunge, rolling 180 degrees and recovering on a heading away from Midway.

I knew the Zeros would stay with the bombers, so I headed away from the island, figuring to climb back up and maybe get a shot at one or two of them a bit later. After climbing to 20,000 feet, I returned to the combat area and looked around. I spotted several enemy aircraft, widely dispersed, giving no sign of mutual support. Although I was alone I still meant to pick off one of the singles.

The metallic resonance of bullets striking your airplane cannot be mistaken for anything else. I realized a Zero pilot had intentions identical to my own, and I wracked the F4F into a tight turn. That momentarily spoiled the Japanese pilot's tracking, but he cut the corner and remained in a position to shoot. A couple of circuits convinced me that I couldn't evade much longer—the Zero would have me for breakfast at that rate.

I rolled my Grumman into another dive, heading vertically into a convenient cloud. I figured the Zero probably would follow, so I chopped my throttle, stomped one rudder pedal, and pulled the stick into the opposite corner of the cockpit.

The violent, uncoordinated movement threw my F4F into a skid. Decelerating and swerving out of plane, I evaded my pursuer, who dived straight past. In a few seconds I had gained the initiative and quickly positioned my gunsight for a shot. However, I had to push forward on the stick to bring my guns to bear and there wasn't time to take off the negative G. When I pressed the trigger, all four guns jammed. The ammunition belts were tossed to the top of their cans, forcing a feed jam.

The Zero disappeared to seek other sport. I was disappointed at losing the kill but happy to remain intact. I tugged on the charging handles for each Browning machine gun and shortly had three .50 calibers working, but the fourth stubbornly refused to function.

Glancing around to take stock, I noticed Midway had taken a pounding. Rolling black smoke rose from the atoll—proof that the raiders had found their targets. But some of them remained in the area. Three widely separated Zeros were visible, apparently freelancing to cover their bombers' withdrawal. That suited me; I could take on one without risk from the others.

The first man I ever killed never knew I was there. A petty officer pilot from the carrier *Kaga* was seeking further opportunity for combat. He had reason for optimism, as the Zeros had just shredded an entire American fighter squadron, claiming forty kills from the twenty-five planes we launched. Completely unaware of me, he probably was looking down when I dived on him from above and behind.

I waited until the Zero's wingspan filled the mil rings of my gunsight. Then, at boresight range, I began firing. Despite the asymmetrical recoil resulting from only three guns, I managed to keep my sight on target and closed in. The fighter took a concentrated cluster of .50-caliber hits and dropped into a spin. It never recovered, and neither of the two nearby Japanese pilots noticed the event. It's likely that the petty officer was not missed until his squadron counted noses back aboard their ship.

That was the end of my shooting. Moments later I heard Midway order VMF-221 to return and land by divisions. Nobody responded. The air operations officer then instructed all pilots to land individually.

Twenty-five fighters had taken off barely an hour before. Ten returned.

Guided to my revetment, I shut down and unstrapped. Climbing from the cockpit, I counted eight bullet holes. Every returning fighter had evidence of combat except one. The undamaged Brewster caused some speculation about its pilot's actions.

My Wildcat and the undamaged Buffalo were quickly refueled and rearmed, but Midway was still reeling from the attack. Gasoline fires raged out of control, hangars were flattened, and debris was strewn everywhere. An Aichi dive bomber had scored a hit on -221's arming area, detonating several small bombs and 10,000 rounds of .50-caliber ammo. Four men had been killed there.

Most of the returning pilots were stunned by their experience. They asked one another what they'd seen and exchanged information with ground crewmen. The CO and exec both were missing; nobody seemed to know if any of the others might have bailed out.

About an hour later, a second alert sounded. Another raid was reported inbound.

My fourth gun still was jammed and none of the bullet holes had been patched, but I had no choice. I climbed in, started up, and began to taxi out. Several F2As could have been flown but nobody was manning up. Glancing back, I saw Capt. William Humberd following in the undamaged Brewster. His own Buffalo had been shot up during the first mission, but he had claimed two kills. I had been two months behind Bill at Pensacola; he was a real nice guy and I trusted him.

Cranking up my wheels, I briefly pondered the odds against two fighters tackling another 100-plane raid. Then I pushed the thought aside. At times like that it helps to have a short memory.

I had already decided that if another raid hit Midway, the atoll would be lost. I planned to ditch at Kure or French Frigate Shoals and hope for the best. However, a few minutes after takeoff Bill Humberd and I were recalled—it was a false alarm. The "raid" was a squadron of SBDs off the *Hornet* (CV-8) returning from a failed search for the Japanese carriers. But clearly Midway was out of the battle. Heading back to base, I consoled myself with the knowledge that Bill had stuck with me. At least someone still had some fight left in him.

Not so most of the others. VMF-221 was a shattered command. Later that day the survivors pieced together what had happened. Major Parks had been shot down near the reef, where he bailed out. The Japanese had strafed him in the water and killed him. Of the other fourteen missing pilots, only one was recovered. Total claims came to eleven shootdowns by -221, including four Zeros. Actual enemy losses to all causes were seven planes. Two were Zeros: one which crashed on Midway and the *Kaga* fighter last seen near the target. Japanese action reports identified the latter victim as Petty Officer Hiroyoshi Ito, who had graduated from flight training in 1939. His identity seems certain, for no other combats occurred at that time and place.

Our squadron fell apart. The senior surviving officer went to the first sergeant and asked if the NCO could run the outfit for the next few days. The career marine replied, "Yes, sir," as expected. With that, the senior captain walked out of the command post, went to a bomb shelter, and proceeded to get drunk. He had plenty of company.

Outwardly I suppose I gave little or no indication of emotion following this calamity, but one entry in my diary two days later gives some hint: "Two fighters still in commission. Feeling better—ready for another fight. Zeros too fast for our planes."

It was a while before we learned of the events at sea from June fourth through seventh. What we found out did nothing to encourage us, for the Marine Corps dive bombers and Navy torpedo planes that took off the first morning were hit nearly as hard as our fighters. The scout-bomber squadron, VMSB-241, lost twelve of its twenty-seven SBDs and SB2Us. Two of the Army's four B-26s returned, but Torpedo Squadron 8's land-based detachment got back only one of its six TBFs. The pilot was Ens. Bert Earnest, with whom I would fly at the Naval Air Test Center three years later.

The remaining days at Midway passed uneventfully. Some of us spent time on Sand Island, where PBYs were bringing back Navy aircrews rescued from the ocean. The Catalinas and Army B-17s maintained long-range searches, but most men relaxed. Following the destruction of four enemy carriers, there was widespread confidence that "the Japs aren't going to bother us."

I indulged in some unexpected socializing. I ran into an old roommate from Pensacola, Bill Hardaker, who gave me a ride back to Hawaii in his PBY. I also met a pair of senior Marine Corps officers out to look things over. One, Col. Claude Larkin, was, like me, from Oregon. He was an immensely likeable officer known as "Sheriff" for his western background and stint in the Horse Marines. I would see a lot of him later. The other, Larkin's aide, was a major named James Roosevelt.

On 12 June I started to pack, and nine days later I left Midway. Riding a PBY-5A on the long flight to Oahu, I had time to speculate on my next assignment. As yet I'd never heard of an island called Guadalcanal.

CHAPTER FOUR

Guadalcanal

Upon arrival in Hawaii I checked in to the Royal Hawaiian with some of the other survivors of VMF-221. Following a few days of relaxation, I reported to MAG-23 for reassignment at MCAS Ewa. After the trauma of Midway it was difficult for some to readjust to the relatively easy life on Oahu, but there were occasional lessons to be learned nevertheless. Passing Bellows Field, I saw a Curtiss P-36 making a loop to a landing—a foolish stunt that obviously had been started too low. The Army plane ran out of room at the bottom of its loop and dived right into the end of the runway, exploding on impact. The P-36 wasn't much of a fighter, but in June 1942 the United States needed every plane it had. The loss of a pilot and aircraft under those circumstances constituted one of the most stupid maneuvers I had ever seen.

Fortunately, I found a far more professional attitude in my new squadron. VMF-223 had been established at MCAS Ewa on 1 May, and was commanded by Capt. John L. Smith, an ambitious, no-nonsense officer. It was peculiar how he was usually called by his full name, like J. Edgar Hoover, although some of us called him "Smitty." But whatever anyone called the CO, or whatever they thought of him, one thing became certain: John L. Smith may have had his faults, but combat leadership was not among them. War correspondent Richard Tregaskis, whom I got to know well in later years, described Smith as "a prairie type; tanned face, wide cheekbones, erect head, a thick neck set on square shoulders and a big sinewey body. He has the steadiest eyes [I] had ever seen: brown and wideset."

Smith's exec was Capt. Rivers Morrel, formerly an Annapolis football standout, and I became the engineering officer. We had a handful of other Midway survivors, including Roy Corry, whom I considered the best of the second lieutenants. I liked him, and his two victories from Midway inspired confidence. Ken Frazier, a twenty-two-year-old pilot from New Jersey, was the most aggressive of the new arrivals.

On 3 July VMF-223 and its sister squadron, -224 under Maj. Bob Galer, began receiving F4F-4s with six guns instead of four. However, the extra guns weren't necessarily an advantage. Although the -4 carried less ammunition than the F4F-3—not a major consideration to me, since I never fired empty in combat—the new Wildcat weighed some 500 pounds more with no increase in power, so it was considerably slower.

A week later we started field carrier landing practice. Bob was one of the few Marine Corps landing signal officers, and he got us ready in his quiet, efficient way. Several days later we flew out to the *Hornet* for our carrier qualifications. The requirement was three launches and three arrested landings, and I made my three "traps" without any waveoffs.

At the time we didn't know where we might be sent, but it was obvious that something was brewing. Therefore, in the limited time available, Smitty drilled -223 on the basics, tactics and gunnery. It was sound procedure but not without cost. On the thirteenth I was practicing dogfighting with Lieutenant Chambers when he apparently blacked out trying to follow me. He dove straight in from 12,000 feet.

Our recent "car-quals" were put to use late that month when the escort carrier USS *Long Island* (CVE-1) ferried fifteen Wildcats to Palmyra Atoll, some 1,000 miles south of Hawaii. On the twenty-fifth we launched all fifteen F4Fs, though one cracked up landing ashore. Thirteen of us then flew obsolescent Brewsters—like those mauled at Midway—back to the ship. Eleven got aboard safely and I took a transport plane back to Ewa the next day.

There was plenty to do upon my return. As engineering officer I was kept busy working out the inevitable bugs in the new fighters. I also selected one for myself—Bureau Number 02100, with a black number thirteen painted on the fuselage. On 1 August I ferried it to Ford Island where VMF-223 was loaded aboard the *Long Island* the next day. It was a frantic, short-notice embarkation but somehow we all made it: personnel, aircraft, equipment, and spare parts. We stashed our gear wherever we could, even in the SBD-3 Dauntlesses of Lt. Col. Richard C. Mangrum's VMSB-232. I didn't know Dick very well at the time, but I got to know him much better where we were going. He was among the finest officers the Marine Corps ever produced—"a Marine's Marine" was how many described him—and a real gentleman besides.

With nineteen fighters and fifteen dive-bombers, the little carrier was crammed to capacity. We manhandled our F4Fs around the deck, spotted them to fire over the side to test the guns, and finally combat-loaded them. The *Long Island* stopped briefly in the Fijis and Efate, New Hebrides, while southbound, allowing brief liberty en route. At Efate I saw some old squadronmates: Chick Quilter, Fritz Payne, and Joe Bauer, now a major commanding VMF-212. We sailed from Efate 18 August, having added an experienced pilot, Loren "Doc" Everton, to VMF-223. Our destination was Guadalcanal, largest of the southern Solomon Islands.

A few pilots began brooding about what lay ahead, and it seemed that those who worried the most were inevitably the ones killed. After Midway I figured I had been through the worst and frankly didn't give much thought to what might happen. I was more concerned with doing a good job, shooting down as many Japanese planes as possible, and watching out for the newer pilots. I had either a sense of fatalism or a certain lack of imagination—I'm not sure which.

We knew very little about conditions on "the Canal." The 1st Marine Division had landed there and on nearby Tulagi on 7 August in the first American offensive of the war, mainly to secure the airfield the Japanese were building. The Japanese abandoned the field when the marines landed ashore, and Navy and Marine Corps construction crews—often using captured enemy equipment—had gotten the strip into acceptable shape by the twentieth.

Launching from the *Long Island* on the twentieth was a tedious process with only one catapult and a crowded deck. Our formation was led by Lt. Col. Charles Fike, executive officer of MAG-23. He wanted his plane spotted last on the deck so he would have maximum possible room for launch, but the carrier's skipper, Captain Burns, had other ideas. Burns, figuring that leaders should lead, ordered Fike's plane spotted first. Strapped into our cockpits, we watched Fike manage a shaky launch. Everybody else got off safely and formed up overhead the ship before heading out.

A few minutes later it was apparent that Fike was headed for somewhere other than Guadalcanal. John L. Smith assumed the lead and got us on course, and we reached the island after a seventy-five-minute flight. There wasn't much time for ceremony after we landed, though Brigadier General Vandegrift and his 1st Division clearly were pleased to see us. Rear Admiral Fletcher, who had led the Wake Island relief force back in December, had pulled his three carriers out of Guadalcanal shortly after the first landings there on the seventh, leaving the amphibious support ships no choice but to hoist anchor and follow. This was understandable from the amphibs' viewpoint since they lacked air cover, but most of the 1st Division's gear had remained aboard the support ships. Consequently, Vandegrift's troops lacked heavy equipment and hadn't seen an American aircraft in weeks, while the Japanese were regularly overhead.

As soon as we landed we began to refuel our aircraft by hand pumps from fifty-five-gallon drums. We learned that the captured Japanese airstrip had been named Henderson Field in honor of Maj. Lofton R. Henderson—better known as Joe—who had been killed in a dive-bomber at Midway.

Guadalcanal's code name was "Cactus," an appropriate word because the narrow beachhead lacked even rudimentary facilities but featured tall, sharp kunai grass on the island's plain. Perhaps the ultimate comment on our crude facilities came from Dick Mangrum: "The general concept of Marine Corps operations envisions rough field conditions, but just how rough is sometimes a bit shocking even to marines!" We were quickly briefed on the local situation and given two basic rules: never go into the jungle alone, and eat and sleep at every chance. Both proved practical.

As deplorable as living conditions were at Henderson Field, we could only guess what they were like for the infantry. It was widely held that Guadalcanal was the only place on earth where you could stand up to your knees in mud and still get dust in your eyes. The island is slightly below the equator, and the temperature and humidity were oppressive during daytime, but nights could be cold. I was ashore three days before I took a bath in the river, or even washed my hands. Our personal gear caught up with us at about that time, so I had my own bedding and my parachute bag stuffed with a change of clothes. Later we became more civilized, and had canvas tents with mud floors.

However, we had no tents that first night, just a canvas fly and Japanese blankets—even Japanese food. Early in the morning of the twenty-first we were awakened by heavy gunfire from the perimeter, about a mile and a half away. As the shooting dragged on we could see tracers slicing through the dark, and I began to wonder what I was doing there, fully exposed in the open under a fly, armed only with my .45-caliber pistol. I remember thinking, "I sure hope those guys on the line can hold them."

But hold them they did. After dawn we learned that a Japanese force of battalion strength had attacked our lines at what we thought was the Tenaru River. Actually, our maps were inaccurate; the Tenaru was farther down the coast, but between 700 and 800 Japanese bodies were counted when it was all over.

Our first missions were flown later that day when Smitty's four-plane division was jumped by six Zeros. All four F4Fs were shot up, and although Smitty tagged a Zero, Tech Sgt. J. D. Lindley crash-landed in number fourteen. That same day we got five more pilots from Joe Bauer's squadron. We needed the help; most of us flew at least one two-hour patrol every day.

A Japanese submarine lobbed some shells into our perimeter in the predawn hours of the twenty-fourth, and a bombing raid arrived that afternoon. It was the squadron's first big combat, and we scrambled fourteen Wildcats to intercept two separate enemy formations totaling about forty fighters and bombers. I was over the water well north of the field when I glanced down and saw a formation of Japanese bombers. It was almost like the Midway setup, but without escorting Zeros. I rolled into an overhead pass and splashed one bomber. Then things fell apart. My division split up and I lost contact with the other three pilots, but we continued to hammer away at the bombers.

This fight was very confusing because it involved twenty-one planes from the Japanese carrier *Ryujo* plus some twenty twin-engine Bettys from Rabaul. It's almost impossible to sort out which enemy formation we hit first. I dropped a second bomber in another overhead run and shot a Zero off Lindley's tail. As the fight drifted toward Henderson I claimed another bomber that was confirmed a bit later. These four kills made me an ace, the first in Marine Corps history, but that thought didn't occur to me at the time. We were far too busy and more concerned about our losses.

In all we claimed twenty confirmed, though I'm told available Japanese records indicate we may have bagged only about a dozen. The *Ryujo* was sunk by squadrons from my old home-away-from-home, the *Saratoga*, later in the day. However, Roy Corry was last seen engaged with some Zeros and Fred Gutt was wounded. And that wasn't all. When we landed and counted noses, we came up three more short: Elwood Bailey, Larry Taylor, and Bob Reed. Fortunately, Reed turned up safe at Tulagi a few days later.

The pattern began to repeat itself: nocturnal shellings and bombings, followed by daytime air attacks. We maintained standing patrols but couldn't always intercept in time, which is what happened the twenty-fifth. An estimated twenty-one bombers, plus fighters, came over at 23,000 feet, but we lacked radar, and the coast watchers, later so splendidly effective, weren't in position yet. I logged three frustrating hours on that mission without firing a round.

Part of the problem was the Wildcat. Combat loaded, it took about forty-five minutes to reach 30,000 feet, so we had to maintain standing patrols. That in turn meant we couldn't meet most raids in force because we couldn't keep enough fighters airborne simultaneously. Things worked better on the twenty-sixth, however, when we tangled with about sixteen Bettys and a dozen Zeros. Smitty got two bombers, I splashed a Zero from a high-side run, and three others claimed kills before the fight broke up.

I was feeling pretty good about the situation as I entered the pattern and lowered my wheels. Suddenly I was jumped by an audacious Zero pilot who apparently had trailed me back to the field. I dived for the nearest antiaircraft gun position, which opened fire and drove off the Zero. Meanwhile I was busily cranking up my landing gear, shoving on full throttle. An airfield where you have to shoot down the enemy just so you can land—what more could a fighter pilot ask for?

There wasn't much hope of catching the Zero, since he was faster and had a head start besides, but as we approached the coast I saw his wing come down and, sure enough, he turned back into me. He was eager to fight. Just over the beach we bored at each other head-on, but I wanted to hold my fire until I was sure he was within range. Then he pulled almost straight up in that startling climb that only Zeros could perform. I had no choice but to try and match him; otherwise he would have a decisive altitude advantage. It had to be a snapshot, as an F4F couldn't climb with a Zero.

At nearly full deflection I got my lead and fired. The Zero blew up, raining pieces down on the beach. I nosed over, checked the area and headed back for Henderson.

There was no doubt about confirmation—several hundred or more marines had seen the whole episode. And I had learned an invaluable lesson: never let your guard down, even over your home field. Later somebody gave me the Zero's oxygen bottle, which had washed ashore.

That was our seventh day on Guadalcanal. I had flown six missions totaling thirteen hours and claimed six victories in three combats.

We received reinforcements fairly regularly, both replacement pilots from other Marine Corps squadrons and Army P-400 Airacobras. The latter were export versions of Bell's P-39, and they were completely outclassed in the air. They didn't have a chance against the Zeros, nor could they get to altitude in time to intercept the bombers. Consequently, they were most profitably used in strafing and light bombing.

Two decades later, in Vietnam, I heard the phrase "Who owns the night?" used frequently. If successful night flying was an indication of who was winning in Guadalcanal, it definitely wasn't our side. Without effective night fighters, we had no way to stop the Japanese from harassing us after sunset. I flew one night patrol from Guadalcanal and accomplished nothing. On the twenty-ninth we had another heckler raid at 0425, with a half dozen or more bombs dropped.

But in daylight we were holding our own. At 1145 that morning another eighteen bombers and twelve Zeros came over. We got ten fighters into the air and claimed eleven kills. Smitty, always an up-front leader, got two bombers, thereby becoming an ace himself, and I claimed one. However, the survivors got through to bomb Henderson, and two hangars were set afire.

On the thirtieth I flew twice for a total of three and one-half hours. The first mission turned into a regular dogfight as I led eight F4Fs into a pack of Zeros that were tangling with some P-400s. I picked off three and Smitty got four among the fourteen credited, but the Army lost three pilots and four planes. We got off lightly, with Lt. Charles Kendrick slightly wounded and the F4Fs of Lts. Ken Frazier and Zen Pond badly shot up.

That same day Bob Galer brought VMF-224 into Henderson, and another SBD squadron arrived with him. We fighter pilots were largely tied to base because of our short range and the urgent need to protect the field, so the SBDs usually operated alone. However, they were making their presence felt, as Dick Mangrum and VMSB-232 had proven.

Much criticism has been leveled at the shoestring nature of the Cactus Air Force, but when I became a general in a later war I looked back on Guadalcanal almost with nostalgia. The differences between 1942 and 1965 hardly could be greater, as in Vietnam we usually had more than enough food, fuel, and ammunition, and plenty of supervision, too. But in some respects Guadalcanal was a pretty nice way to fight a war. We were pretty much on our own, and that counts for a lot to those doing the shooting.

As for morale, it wasn't a problem. I don't recall any serious doubts that we could hold Guadalcanal. Had worse come to worst, the aviators could have evacuated on short notice, in much the same way that I contemplated ditching at an outlying island had the Japanese occupied Midway. It would have been a grim scene for everybody else, however.

There was only scattered action during the next week, but I missed even that. Like most of us on Guadalcanal, I came down with dysentery, and I didn't fly from 31 August through 3 September. However, my Cushman scooter had arrived and I got it running again. A few days later my Zenith radio was in operation, too. Most pilots spent a lot of time playing poker on an improvised table in the ready tent while waiting for the Japanese phone to jangle in an alert. Little diversions like these were important, as they helped break the monotony just as they had on Midway. Fatigue was always a factor in attrition, but with my diversions and my ability to sleep at any time, I was better off than most other pilots.

At 0100 on the fifth, three enemy destroyers began shelling our beachhead, and the U.S. destroyer-transports *Gregory* (APD-3) and *Little* (APD-4) were sunk while trying to intervene. Shortly before noon—the usual time, allowing transit south from their bases at Rabaul, New Britain—Japanese bombers were reported inbound. Elements of both of our fighter squadrons intercepted, claiming seven confirmed Bettys and one probable, but our exec, Rivers Morrel, was wounded and out of action, and -224 lost two pilots. Unable to get above the Zero escort, we had been caught at a disadvantage.

The night of 7–8 September was something of a disaster for us, at no expense to the Japanese. We put up several F4Fs to stop the harassing attacks by enemy aircraft that deprived so many marines of badly needed rest. There was no contact, but one Wildcat cracked up on takeoff and four more on landing. This was probably the last attempt at launching night fighters until the Army's P-38s arrived.

The next day was one of mixed blessings. VMF-223 and -224 intercepted a raid of thirty-six bombers attacking a convoy in the channel between Guadalcanal and Tulagi. The Bettys had fighter escort, but I got two. That was the good news. Maybe a numerologist could have predicted what happened next, as I was flying my thirteenth mission at Guadalcanal, in my old number thirteen Wildcat, and had just made my thirteenth kill. Next thing I knew, I was sitting in a flying junk heap with a fire in the cockpit. Some crafty Zero pilot had hit me before I even knew he was there. I had no choice—I slid back the canopy, undid my safety harness, and went over the side at 22,000 feet.

Fortunately, I had a good chute and made a safe landing in the water about 400 yards off Neil Island, thirty miles from base. My only injury was a rope burn where the parachute risers had chafed my neck. I didn't know it at the time, but Clayton Canfield was also shot down in that fight. He splashed a Betty before bailing out, but he landed alongside an American destroyer and returned to Henderson. It seemed oddly consistent, as we had been through a similar experience together at Midway.

I bobbed in the water for about four hours before a native named Steven rowed out. He had a tough time paddling against the tide that last 200 feet, but he fished me out and gave me some coconut milk. Then he delivered me ashore on Guadalcanal—still about twenty miles from Henderson Field—to a Fijian doctor named Eroni who spoke fluent English and took excellent care of me.

I was stiff and sore from the bailout, so I rested on the tenth. I learned that Japanese were in the vicinity. The next day, led by Eroni and several other natives, I started back to Henderson Field. We reached Parapis, where there was a radio at the former district officer's place. I think he had abandoned the office, but the radio remained. I could only get it to work on the receiver side, so I gave up and settled down to dinner of broiled duck and eggs.

On the morning of the twelfth we continued our trek to Henderson but had to turn back. Natives were evacuating the area because an estimated two thousand Japanese now stood between us and Henderson. Upon returning to Eroni's little camp, I went to work on his eighteen-foot motor launch. It had a five-horsepower single-cylinder engine that hadn't run for quite some time.

We planned to leave at 0330 on the thirteenth, but I couldn't get the engine to start. I spent most of the day tinkering with it, and that evening we ate the last of Eroni's chickens. The Japanese had gotten the rest in our absence. Next morning, again at 0330, we set out with the single-banger putt-putting away and arrived at dawn. Late that afternoon we had the pleasure of watching some Wildcats knock down three Japanese floatplanes, but it turned out we had only seen half the action. Maj. John Dobbin and Lt. George Hollowell of Bob Galer's squadron were credited with six shoot-downs in that episode.

My return from the dead caused something of a sensation. A VMSB-232 enlisted man, Dennis Byrd, told me years later that "gloom settled over the island" when I failed to return. It's a fine compliment, but I was too preoccupied to notice it at the time. I spent most of the next day collecting my gear that had been redistributed and kicked a Navy pilot, Wally Clarke, out of my bunk. However, I never did get back my shower shoes. An infantry major named Shoup had happened by shortly after I was reported missing, and needed a bunk for the night. He had lost his tent in a bombardment, along with most of his gear. When he left he took my shower shoes with him. Twenty years later I learned the rest of the story from Commandant Gen. David M. Shoup—who was still reluctant to return my shower shoes.

Besides scrounging for gear, I helped Eroni pull his launch out of the water for some painting. He had been awfully good to me, and I made sure he was well treated during the several days he stayed with us.

On the nineteenth I had an unexpected reunion with a college classmate. Capt. John Wilkins, an Army fighter pilot, temporarily joined VMF-223. He flew a few times, I think, and though there wasn't any activity during that period it's notable that we needed pilots so badly that an Army aviator was incorporated into a Marine Corps squadron, in an aircraft he had never flown! That same day I had an inconclusive combat, and no losses occurred on either side.

John's appearance in -223 wasn't quite as much of a fluke as it may seem. Military aviation still was a relatively small community in 1942, as very few wartime-trained pilots had reached combat units. I'm reminded of three other Oregonians, graduates of Linfield College near Yamhill. Two of them became Marine Corps aviators: Ken Jernstedt went into fighters and eventually joined Claire Chennault's Flying Tigers, and Bruce Porter flew dive-bombers at Midway and was Dick Mangrum's ops officer on Guadalcanal. The third, Rex Barber, became an Army pilot, and in April 1943 he flew the famous Admiral Yamamoto interception. There's no longer much doubt that Rex shot down the admiral's plane on that mission.

During the lull I picked out a new Wildcat, Bureau Number 03508, and immediately had a number thirteen painted on it. Darned if I was going to let superstition get the better of me. Eroni left on the twenty-fourth, and during a scheduled patrol I saw his launch being towed by a Higgins Boat back to Neil Island. Next day I flew over his place and dropped a package with some items I knew he could use.

In the five days it had taken me to get back to Henderson—from the tenth to the fourteenth—VMF-223 had claimed twenty-four more victories. Five of those went to Smitty, now a major, who had raised his score to sixteen. It has been reported that I no sooner reached Henderson than I looked up Brig. Gen. Roy Geiger, commanding 1st Marine Aircraft Wing and senior Marine Corps aviator on Guadalcanal, and insisted that Smith be grounded for a similar period so I could catch up. That's not exactly how I remember it, but the story does illustrate the feelings of rivalry between fighter aces. In retrospect I wonder if getting shot down cost me the leading Marine Corps (and therefore, U.S.) score at the time, and maybe the Medal of Honor. But at least I was alive, so I can't complain too much!

On the twenty-seventh there was another raid and plenty of shooting. Between 1330 and 1445, elements of both F4F squadrons intercepted the Japanese almost directly over Henderson Field. I took Smitty's airplane because he had the day off and mine was down for maintenance. I latched onto Maj. Kirk Armistead's division from VMF-224—he was another Midway survivor of VMF-221—and we split a Betty between us. I also bagged one on my own, narrowing some of the gap on Smith. This mission restored whatever confidence I may have lost as a result of being shot down, but frankly that hadn't been much. I learned early in my career that if I kept my eyes moving I would be all right in a dogfight. I was blessed with the ability to look at another airplane and instantly know what it could do relative to my speed and position. I rarely made conscious decisions in a combat—mostly I acted on instinct.

There was another large raid at 1300 the next day, with Navy F4Fs from Fighting 5 participating as well as VMF-223 and -224. There were about twenty-seven Japanese planes, and the three F4F squadrons claimed most of them destroyed. I got one, and my diary notation says, "We really cleaned them out, 23 of 26 or 27."

Flying was canceled on the thirtieth because of rain, but Adm.Chester Nimitz, commanding the Pacific forces, arrived on an inspection trip. I had only seen him once before, at Midway, but got a chance to talk to him the next day. I really admired him—there weren't many four-stars who got within shooting distance of the enemy, and his concern for his people was genuine. When he decorated Smitty, Bob Galer, and me with the Navy

Cross, our aggregate score at that time was about forty-three enemy aircraft, so no doubt the medals were deserved. I've seen film clips of that event through the years, and I'm struck with the informality of it all. The three of us are dressed in crumpled khakis and wearing dark-blue baseball caps—mine with the brim turned up—and .45s in shoulder holsters. That was sort of the uniform of the day on Guadalcanal; functional if not stylish.

By now we had settled into something of a routine. At night we'd sit in front of our tents, some of us dressed in pajamas as if that were normal evening attire on Guadalcanal. Liquor and chocolate were strenuously rationed, and although the pilots got most of the booze for "medicinal" purposes it was interesting to watch a candy bar divided into 160 parts for all hands. I spent a lot of time writing letters, even though we didn't receive much mail. Occasionally somebody would crank up the portable phonograph, though more often there was improvised music. One popular addition to "The Marines' Hymn" was composed on Guadalcanal, and it went something like this:

> When Gabriel toots his mighty flute,
> calling old campaigners home,
> and when Tojo's balls hang from the walls
> of Valhalla's golden dome,
> then the Lord will look at Vandegrift
> who is eating spam and beans,
> saying, "God on high sees eye to eye
> with United States Marines."

Things were back to normal 2 October, and not for the better. Smitty led a hasty scramble, but we couldn't get off quickly enough and I didn't even get a shot. Smith and Galer each scored but both force-landed badly damaged aircraft, and we lost Willis Lees and Charles Kendricks besides.

The next day I had one of the most satisfying combats of my career, and not merely for personal reasons. Joe Bauer, now a lieutenant colonel, had come up from Efate and asked Smitty if he could fly with -223. Smith said, "I'm not the flight leader today. You'll have to ask Marion." So Joe asked if I minded if he would fly in event of a raid and I said, "Sure, you can take the second section." As I've noted, my professional relationship with Joe started poorly but eventually warmed to one of mutual regard and respect, especially in the prewar days at San Diego. I knew him as a highly proficient pilot, a real hard charger.

As luck had it, we were scrambled during the noon hour but didn't see anything for a while. So we kept climbing and reached 30,500 feet indicated, which I computed as nearly 35,000 actual. It was the highest I had been in any aircraft up to that time. Shortly before 1300 I looked down and noticed several Zeros way below, at about 12,000 feet, northwest of the field. I had a lieutenant colonel back there and I was only a captain, so I came up on the radio: "Look, we have about ten Zeros down there. Are you ready to go?" I got no answer and decided Joe's radio wasn't working, which in fact proved to be the case. So down we went.

I tailed in behind the last Zero on the right and destroyed it with a thirty-degree deflection shot from about 100 yards. Another came up from below me, and as I pushed over I pulled the trigger and all six guns quit. That's just what had happened to me at Midway. The ammo immediately jammed in the chutes under negative G, and by the time I got a couple of guns cleared the fight was over.

When we landed, Joe Bauer was jumping up and down. He said he got four definite and one probable which, combined with the bomber he had bagged a few days previously with -224, made him an ace. Lt. Ken Frazier claimed two and then was shot down. While parachuting toward the water a Zero shot at him but fortunately missed, and Ken was picked up by a destroyer. He'd done exceptionally well at Guadalcanal, finishing as the third-ranking ace in the squadron. Lt. Conrad Winter also scored, so we totaled nine confirmed and one probable for one F4F lost.

I asked Joe how his plane had performed, and he said his engine was running pretty rough. I had the plane captain check it, and he said Joe had left the engine in high blower. It's amazing the engine had held together while running at full throttle down around 10,000 to 12,000 feet, but it was Joe's day all around.

Early on 10 October Smitty led an escort for some SBDs and TBFs that ran into floatplane Zeros and biplanes near New Georgia. In a short, sharp combat the eight Wildcats splashed nine floatplanes. I flew twice that day but drew a blank, and I admit to having licked my chops in envy. I knew I could have brought my score up in a hurry if I had been in the right place, since there's nothing as much a sitting duck as a floatplane.

When we left Guadalcanal, Smitty was the top American fighter ace of the war at that time, with nineteen confirmed, while I had sixteen in thirty-two missions involving ten combats. Six others also qualified as aces. Overall, VMF-223 was credited with 110 aerial victories from 20 August to 12 October, but at a price of 57 percent casualties among our own pilots—six killed and six wounded—plus some losses from other squadrons on detached duty. Also, only eight of our original Wildcats remained.

Our war temporarily was over. I looked forward to going home.

CHAPTER FIVE

Edna

John L. Smith, Dick Mangrum, and I landed at Hamilton Army Air Field, near San Francisco, on 22 October 1942. We were ten days out of Guadalcanal, and we had priority air travel orders instead of the liberty ship transportation the other members of our two squadrons were afforded. At first that appeared to be good news, but then we learned the awful truth: we were going on publicity and war bond tours for the Navy and Treasury Departments.

However, I was able to spend a few days in the Bay area, visiting friends and relatives between press conferences. I made a special point of stopping in Lindsey, California, to visit Clayton Canfield, my steady wingman from Midway and Guadalcanal. Then in Santa Ana I looked up Roy Corry's and Walter Swansburger's families, who were anxious to learn something of their loved ones' deaths.

The "rubber chicken" circuit began in earnest on the East Coast in November. Reunited with Dick and Smitty, I submitted to the grind of radio shows and appearances in New York, Washington, and Navy flight schools. Public speaking never has been one of my favorite activities, and of the three of us there wasn't much doubt that Dick was the most effective speaker. He was always articulate, poised, and immaculately dressed, and audiences always liked him.

A couple of worthwhile interludes were sandwiched in between public appearances. I got to fly the F4U-1 Corsair for the first time during this period, and it made a most favorable impression. Smitty and I were scheduled for two talks at Grumman, and on 19 November we checked in to the Waldorf Astoria. I skipped lunch to fly an early F6F Hellcat, and considered it a worthwhile trade. Dick had gone to the Brewster Aircraft plant where they built SBA dive-bombers, and he joined us that evening for a cocktail party sponsored by Grumman.

Smitty and I were Grumman's two best "customers" at that time, and the company spared no expense in demonstrating its gratitude. However, we flinched when *Life* magazine described us as "the heavenly twins." That's the sort of thing that takes months or years to live down. Our outlook quickly improved when we learned that four models from the John Robert Powers agency would be at the party. As the only bachelor, I told Smitty and Dick that I expected to get my choice of companions for the party. They mumbled between themselves but tentatively agreed.

The four models were ready on time, and I quickly singled out a vivacious, raven-haired girl as the most appropriate for me. Her name was Edna Kirvin; she was fun, she was beautiful, and she was nineteen. After cocktails Edna and I moved on to other diversions and ended up spending the rest of the evening together. We started with Fred Waring's radio program, then went to the Twenty-One Club, next moved to the Copacabana, and finally closed up the Stork Club at 0300.

I didn't get Edna home until about 0400. She had only promised to stop by the party for half an hour on her way home, and as the night wore on she kept calling her mother with progress reports. That impressed me. It was obvious she came from a good family, and later I learned that our backgrounds had some similarities. Her father had also died early, leaving her mother to run a trucking business in the rough-and-tumble New York environment.

The rest of the month's schedule took us to Washington, Jacksonville, Miami, Pensacola, Corpus Christi, Iowa City, and Chicago. More fashion-model dates were provided in Chicago, where we did the College Inn and the Latin Quarter. I found myself comparing New York's models to Chicago's, and I much preferred the former. Edna was never far from my mind. While on Guadalcanal I had been writing to several girls in the States, and I looked up three or four of them while cruising around the country, but that ended when I ran into Edna.

On 9 December I was back in New York after flying an SNJ on instruments through a snowstorm en route from Ohio. Following an appearance at the Lynden, New Jersey, factory where General Motors built FM-1 Wildcats, I took Edna on another dusk-to-dawn whirl, this time with Smitty and Ken Frazier.

On the seventeenth I asked Edna to marry me. She said she would have to think it over, but at least she didn't say no. Dick took her to a couple of functions while I was away, and was a perfect escort. He would get her home at a civilized hour and gallantly kiss her hand at the doorway. Maybe it was a good thing he was already married!

While the big question hung in midair I took a train cross-country to Portland. My cousin Wilbur Carl was a dollar-a-year man for the Treasury Department and talked me into a bond tour. As a former sales manager for Fields Chevrolet in Portland, he was a pretty high-powered salesman. So on the twenty-second I returned to Hubbard, where the governor sponsored a banquet for me. Two days later my mother and sister and I visited my brother and his wife at Fort Lewis, Washington. Manton was now a first lieutenant, adjutant to the base brigadier general in the field artillery.

Three days after Christmas I happened to be in Meier and Frank, the largest department store in Portland. Somehow word about Edna had leaked out, and while visiting in Mr. Frank's office he asked me about her. Things quickly got out of hand, and next thing I knew Mr. Frank handed me the phone and I was talking to Edna in New York. She agreed, long-distance, to marry me!

I started planning a return to New York, but there wasn't much time. I had three speeches and a radio show the next day and five appearances the day after. But somehow I got an engagement ring and called Edna again to set the date. We tentatively agreed on 8 January in Brooklyn.

I took the Streamliner eastbound on New Year's Eve and, following a stopover in Chicago with millionaire John Watson, I arrived in New York 4 January 1943. I quickly checked into the Biltmore, and was at the Kirvins' by 0930. There were minor problems, as seems usual for such endeavors. First, the ring was too small and had to be adjusted. Then we got our blood tests and had an interview with her parish priest, Father White. Our religious differences could have been a stumbling block since Edna's family was Catholic, but we decided to skip a church wedding.

Following a hasty trip to Marine Corps Headquarters in Washington, D.C., I was back in New York. Edna and I again met with Father White and then got our marriage license, so things were tracking nicely. Then the press learned I was back in town. It was just the sort of story the wartime papers loved: "Returning Hero to Wed Local Model." I think the *Journal American* broke the story, and we were a little concerned about other news hounds jumping on it. Edna and I were married in Brooklyn at 1745 on the eighth of January 1943. There was a reception at the Towers Hotel before we returned to the Biltmore.

It was a long, long way from Guadalcanal.

At first I thought Edna and I might have more privacy once we got to Oregon, but no such luck. *Life* put a writer and photographer on our tail almost as soon as we arrived in Portland on the twelfth, and local reporters also joined the hunt. That day I was scheduled to address the legislature in Salem and had a two-hour appearance at Oregon State College in Corvallis. The next day was occupied with a radio broadcast and an appearance at Grant High School in Portland. They kind of mobbed us, both at Oregon State and Grant, although I got the impression that the boys were a lot more interested in Edna than in listening to me. Some of the college students in particular seemed to think that if they could make out as well as I did, they'd join up the next day.

We all returned to Hubbard, and by "we" I mean Edna, me, and the *Life* team. Their presence created an interesting logistical problem because my mother had only one room available for the female writer and male photographer. But since that was the only place for them to sleep, they made whatever adjustments the journalistic trade required. At any rate, they got their pictures and story.

One aspect of the story was never published, as far as I know. While I was overseas my family's dairy barn had been torched, and some ninety cattle had died in the fire. There was no doubt it was arson, as investigators found where two fires had been started simultaneously. I don't think the perpetrators were ever found, but it wasn't hard to guess who was behind the crime because there were German sympathizers in the area. Because I was fighting Germany's ally, Japan, evidently my family was singled out for special attention. It's hard to imagine that sort of thing occurring in the United States during World War II, but foreign-born U.S. citizens sometimes retained loyalties to their native lands. In retrospect, it wasn't an unprecedented situation—French communists reportedly assisted the German invasion of France in 1940 because Hitler and Stalin had signed a nonaggression pact.

There was a bittersweet result of the barn burning. Local school kids took up a collection to offset part of the loss and tried to present the funds to my mother. However, she was a proud woman and declined the offer, though she really did appreciate the sentiment behind it.

Still a captain, I was to take over VMF-223 at the new air station at El Toro, south of Los Angeles, during late January. But Edna and I took our time and first went to San Diego. We arrived the twentieth and stayed for a few days at the Hotel Del Coronado, an area landmark and an elegant residence. I was able to see Smitty for a day or so as he was preparing to leave for Cherry Point, North Carolina. Edna also got to meet Bob Galer and John Carey, and she was an immediate hit. I realized that she and I were a case of opposites attracting, since she was extroverted and made friends without effort while I usually was more reserved.

Edna certainly was given a quick introduction to the Marine Corps. She was married barely a month and already the wife of a squadron commander. As such she had some social obligations, but she handled these without a hitch. I was really proud of her, and I was thankful for the attention she received from the base commander's family. They could not have been nicer. Meanwhile, we hunted up a two-bedroom house on North Coast Boulevard in Laguna Beach, where we lived for about the next six months.

VMF-223 was in pretty poor shape at the time I inherited it. All the Guadalcanal pilots had been reassigned except Ken Frazier and Fred Gutt, and we had no aircraft. I flew an occasional SNJ or F4F just to keep current, but it was mid-March before we scared up a ragtag assortment of eight Wildcats and SNJs.

Around this time, I undertook an endeavor perhaps more exciting than flying fighters—teaching Edna to drive. Fortunately (or unfortunately, as the case may be) I was well-known to the local police; in six months I had professional dealings with more than one of them. Eventually Edna was safe to solo, even with the three-speed transmission in our Chevy sedan.

That spring was a kaleidoscope of activity. At the dedication of El Toro Marine Corps Air Station on 17 March a pilot was killed while slow-rolling an F4F in front of the crowd—and his wife. I got a second Navy Cross and a promotion to major on 3 May, and Ken Frazier married a local girl named Connie. I flew to Camp Pendleton for the filming of the movie *Guadalcanal Diary*, based on Richard Tregaskis's best-selling book. The exact dialogue eludes me, but my screen career lasted approximately five seconds and consisted of something like, "They've stopped giving out fuel with eyedroppers."

In early April, while flying an FM-1 Wildcat, I tangled with some P-38 Lightnings from Orange County Airport—now John Wayne Airport. We had an exchange program and knew one another to a degree, so we agreed to jump each other whenever appropriate. Also, I visited John Wilkins, my classmate from Oregon State and onetime squadronmate from Guadalcanal. He was now a major commanding a Lightning squadron at North Island, San Diego.

John was having trouble teaching his pilots overhead runs and asked if I could help. During the few days he had been with us at Henderson Field he had been impressed with the overheads we made on Japanese bombers because of their safety for the pilot and their effectiveness against the enemy. An overhead run accomplished two important things for a fighter pilot: first, it largely eliminated bullet trajectory from the gunnery equation because you were shooting straight down toward the target; and second, the run denied enemy gunners a shot at you because you were diving on them from directly overhead. True, a bomber's tailgunner could snipe at you as you flashed vertically past him, but if you did it right he was dead by then.

However, I discovered there was a lot of difference between the F4F and P-38. The first time I made an overhead in a Lightning I really thought I would buy the farm because I began at the same altitude I had used in the

F4F, causing me to build up way too much speed. The P-38 was a sleek, fast aircraft, one of the first ever to encounter compressibility. I flashed past the target and had trouble pulling out. Every time I pulled back on the yoke I got into buffet, and when I eased off the airplane went faster. It was a pretty rough ride.

I kept working at it and eventually made a pretty good pass, but I had to slow down close to the stall and begin my pullout before I piled up much speed. Unfortunately, that early model P-38 did not have speed brakes, and several planes and pilots were lost before the Army got control of the situation. Later that year John took his squadron to the European Theater, where he went missing in action. At war's end he was declared dead, though I never learned any details.

By June VMF-223 had five F4U-1s, the early "birdcage" model with the low, framework canopy and bouncy landing gear. I tangled with a lone P-38 and was beating up on it when I lost part of my elevator at 320 knots indicated. I learned that the Lightning couldn't turn with a Corsair but had more speed and climb. One nice thing: P-38s were all over the place. If you went looking for trouble you could usually find a P-38 somewhere from March Field, Orange County, or North Island. I found I could outturn a Lightning, but it was faster on the level and would outclimb a Corsair.

Most of my pilots had relatively little flying time, and we had plenty of accidents. A senior lieutenant lost it one day during a dogfight, got into an inverted spin at 5,500 feet, and bailed out at 3,300. The F4U righted itself and crashed right-side up east of El Toro—one of life's embarrassing moments. Several Corsairs were lost under similar circumstances, apparently because the training command had stopped teaching inverted-spin recovery. Pilots often didn't recognize that they were in that situation, and if they kept the stick forward, as in a standard recovery, they remained in the spin.

There were also many ground loops on landing. In the birdcage model the pilot sat low in the cockpit and the tailwheel strut was too short. Landing three-point, some pilots would stall too soon, and the aircraft would whip back and forth on rollout. I recommended landing tail-high, holding that attitude until speed fell off and the tail settled naturally. Still, one pilot put an F4U on its back during landing, and until the -1A model arrived with its higher canopy and longer tailwheel strut the problems continued. Ken Frazier and I were all right, and even short pilots like Fred Gutt could handle the -1 if they were careful, but VMF-223 needed a lot more experience.

Ironically, one of my biggest problems was trying to avoid getting someone in the squadron who was senior to Ken or Fred. I wanted to keep them in positions of authority because I knew and trusted them. The group commander, Colonel Freeman, heard about my designs and called me in. He said he was going to assign incoming officers as he thought appropriate, and consequently Frazier and Gutt were pushed down the ladder. That's how things were done in the Marine Corps; experience and ability seldom overcame seniority.

Around 10 July we heard we might leave soon for overseas, and for once the rumors were true. On the eighteenth the entire squadron squeezed into a dirty, crowded troop train and headed north. We boarded the seaplane tender USS *Wright* (AV-1) at Alameda on the twentieth and arrived at Pearl Harbor a week later. Edna, a bride of barely six months, returned to modeling in New York City.

VMF-223 spent the rest of the summer and fall at MCAS Ewa and Midway. We received two majors—Bob Keller and Al Armstrong—from VMF-212 despite the fact that their CO wanted to keep them and I wanted my own senior pilots. Eventually Bob ended up as a lieutenant general and Al made major general.

Midway's facilities were much improved from my previous tour, but that didn't compensate for everything. We lost one pilot and four aircraft while there, including three F4Us with engine failures over water. However, my personal Corsair was number thirteen, and it performed well: 275 knots (316 mph) in a speed run at sea level.

It seemed we continually received pilots with little or no time in the aircraft they would fly in combat. Their total time was very low and some of the young fellows were not enthusiastic about flying fighters. I thought there had to be a better way of screening these pilots and providing them a little more flight time, particularly in unusual maneuvers. Many of them not only weren't very good at aerobatics, they weren't even interested in learning. They tried to stay away from anything that put them on their backs. It took a lot of doing to convince them they had to be able to recover from any position at almost any speed.

I found I could help in some ways. For instance, I told my pilots about the Zero over the beach at Guadalcanal. If you had to shoot with your nose pointing straight up, you could do it with a certain amount of confidence. But I stressed that nobody should try to dogfight a Zero. If you didn't get a shot in the first ninety degrees, reverse the turn and dive away—extend out of range and come back again with position or altitude in your favor. These were some of the points I had to get across as soon as possible. I couldn't reject pilots based on their attitudes, even though attitude had enormous significance in combat flying. If a pilot doesn't want to fly hot airplanes and mix it up, he shouldn't be in fighters.

Before leaving the States we had wisely bought fifty cases of liquor and packed it for the trip to Hawaii and Midway. I wasn't much of a drinker, but

I knew the value of booze as barter. Since we were on Midway for two months, our liquor stash was a source of pilferage and the atoll's commodore threatened to confiscate it. Finally I appealed to his sense of duty and used a whole case to buy him off. Aboard ship we had the liquor under guard, as it became more valuable the farther west we went.

VMF-223 sailed in the escort carrier USS *Breton* (ACV-23) with half of VMF-216, headed south from Pearl Harbor on 30 October. My twenty-eighth birthday was observed at sea two days later, but some friends in Hawaii, the Kepplers, had thrown an early party for me before sailing. I had joked with Edna that she never would have an excuse for forgetting my birthday because her church always observed All Saints' Day the first of November.

On the tenth day at sea we launched fourteen Corsairs for Efate in the New Hebrides. After spending a few weeks reorganizing and letting the rear echelon catch up, we were airlifted to Barakoma Airfield on Vella Lavella Island in the Solomons. Immediately I was concerned with two problems: aircraft and weather.

The Marine Corps had a policy of rotating squadrons in and out of the combat area, leaving the same aircraft for units to fly in rotation. Since nobody really owned the airplanes, there was little motivation to keep them maintained properly. I tried to get this policy changed at group headquarters but made little progress.

November is monsoon weather in the Solomons, with lots of rain and low ceilings, and canceled or diverted missions were not unusual. Consequently, our first mission wasn't flown until 1 December 1943 when six of us strafed Chabais. We lost Lieutenant Kessler, cause unknown, but he returned three days later, reporting his engine had been hit.

Commander Air Forces Solomons (ComAirSols) had a major priority of conquering the Japanese naval-air facility at Rabaul, New Britain. Bomber escort was mandatory owing to the heavy enemy fighter strength at the Rabaul fields, and consequently the Allied fighters staged through advance bases for each mission. Otherwise they'd have lacked adequate range. In mid-December ComAirSols started a concerted effort to gain air superiority over Rabaul with a series of bombing missions and fighter sweeps. Usually VMF-223 and the other F4U squadrons staged through Torokina on Bougainville when weather allowed. My first sweep to Rabaul was flown on the sixteenth with seventy-seven fighters, but we made no contact.

As a squadron commander I had a turn at leading the Rabaul missions, which aggressive aviators regarded as gravy trains. Greg Boyington had VMF-214 at this time, and he was getting score-happy. Joe Foss had completed his Guadalcanal tour in February of that year with twenty-six victories, and reportedly Boyington was closing in. Actually that wasn't the case, as we learned forty years later. When Greg left prior to getting kicked out of the Flying Tigers in 1942, he returned home claiming six kills in China. General Chennault had credited him with only two aerial victories, though for some reason the Marine Corps accepted the higher figure. But in late 1943, with his combined AVG and VMF-214 scores, Boyington was nearing Joe Foss's record as the Black Sheep's tour neared completion.

So, when I reentered the South Pacific "ace race" it was a close contest. I was certainly competitive as a fighter ace, but since I was only recently back in combat I could afford to take it easy at first. On 23 December I led my guys up to Torokina again, where we topped off and launched for Rabaul as I led twenty Corsairs and twenty-eight P-38s. We made contact over Cape St. George and I stalked a new opponent through the clouds. It was a Kawasaki Tony, a sleek, good-looking Japanese Army fighter that resembled the Messerschmitt 109. I splashed him between Rabaul and New Ireland while the rest of the squadron claimed three more confirmed and three probables without loss. Boyington's squadron and VMF-222 added fifteen more; we were taking a toll on Rabaul's defenders.

I flew a routine escort the next day and returned in a sweep on the twenty-seventh. We were up at 0415 to escort a PBY to Torokina, where we refueled and continued to Rabaul. Shortly past 1000 we spotted a flock of Zekes, though I probably shouldn't have pressed it. My radio, airspeed indicator, and altimeter were inoperable but my guns worked, so I went after a pair of Zeros. I killed one and damaged another while VMF-214 and -216 F4Us claimed sixteen more.

One thing quickly became evident on these missions. Although I always led -223, and sometimes the entire fighter sweep, nobody could control so many aircraft simultaneously. It was a flight leader's war, with four-plane divisions usually fighting their own battles in their particular slice of sky. I led another sweep on the thirtieth but got no further than Torokina owing to bad weather. The same thing happened the next day.

On 2 January 1944 Greg Boyington looked me up at Vella Lavella, where I was scheduled to lead the next day's sweep. Greg said he had twenty-five kills but that the Black Sheep's tour would end in a couple of days, so this was about his last chance to break Joe Foss's record. Greg asked if he could take my mission, though in his book he stated that I offered him the chance. At any rate, I thought I was giving up little because -223 still had lots of time left in combat.

Boyington led the mission next day, but he was shot down and captured by the Japanese. Their records show only two planes lost in the air on 3 January (U.S. pilots claimed nine), and Greg had witnesses for only one of the three he later claimed on the mission.

When Greg came out of captivity in 1945 he had been awarded a "posthumous" Medal of Honor and was proclaimed the Marine Corps's leading ace. The medal was deserved—Greg was a talented aviator and an aggressive combat leader—but I've been rankled by the "top gun" title ever since. Even allowing the two unsubstantiated claims from his last mission, he couldn't match Joe Foss's total or his score in Marine Corps service. To my knowledge, Joe never has made any fuss over the situation—he's too much a gentleman for that—but for the Marine Corps officially to recognize Boyington as its top ace, despite documentation to the contrary, defies all logic. I suspect it's a bureaucratic inability to admit such a long-standing error.

Two days after Boyington went missing, the first of VMF-223's three scheduled line periods was over, and the pilots headed for R and R in Australia. The ground echelon remained at Barakoma to prepare new aircraft and support VMF-215. This was another inequity in the system. The enlisted men lived in the same difficult, unsanitary conditions as the pilots but didn't get the mid-tour breaks. However, there was some reason for satisfaction at this time. From 23 December to 4 January we were credited with fifteen confirmed, nine probables, and six damaged, a tally exceeded only by two other Corsair squadrons and a Hellcat outfit. Despite the combat and the miserable flying weather, we had lost just one pilot. It could have been far worse.

Upon return from Australia in early February the roof fell in. I reported to Efate, down in the New Hebrides, and learned I had been transferred to Marine Aircraft Group 12 staff—despite my objections. Bob Keller took over -223 for the rest of its tour. It was the end of my shooting, and the weather reflected my mood—we had hurricane-force winds that toppled a lot of trees.

My time at MAG-12 dragged by. We moved from Efate up to Emirau in the Bismarck Archipelago in May, keeping pace with the war. There were a few diversions, however, including the arrival of VMF-115 with Joe Foss as CO. I hadn't known Joe previously, but we got well acquainted. Medal of Honor winners like Joe weren't expected to return to combat because the Navy didn't like the idea of losing its heroes, as had happened to Butch O'Hare in late 1943. But aggressive aviators like Joe, Zeke Swett, and Ken Walsh managed to get back into action, where their experience was invaluable.

Perhaps the most fascinating event of my staff time came on 29 May when I got to fly with Charles Lindbergh. I thoroughly enjoyed talking to him; he was a real gentleman. And he knew the Corsair intimately, being a field representative for United Aircraft, which owned Vought and Pratt-Whitney.

Our mission wasn't anything significant—a routine bombing hop to Kavieng where there was no aerial opposition. However, Lindbergh went out to his plane well before I did and inspected it to the last rivet. When we started engines he sat there and listened to his R-2800 for the longest time, then taxied back and forth before signaling he was ready to go. Once airborne, it was routine. He flew good formation and made good dives.

Although I flew a few missions in F4Us and even one night hop in a Hellcat, most of my flying was in the staff aircraft. This was a Grumman Goose, a twin-engine amphibian designated JRF in Navy service. Col. Vernon Guymon, the group commander, usually let me use it whenever I wanted to. I checked out some other pilots on water landings but none of them seemed to take much interest, whereas I enjoyed the experience. In September I used my influence and put Edna's brother Bernie on as second crewman on the JRF. The Kirvins were an air-minded family, and her brother Harold had made a career at Grumman.

Late that month I flew over to Mussau Island to pick up an Australian pal of mine, Warrant Officer Bob McKee. I brought him to Emirau so he could visit some friends for about three hours, and when I took off again to return him to his base I realized he was drunk. Not only that, he wanted to take control of the plane. In fact, he insisted on it. What to do?

I had an autopilot in the JRF that I seldom used, so I switched it on and let McKee take the controls. He immediately tried some violent maneuvers; pushing and pulling, fighting that autopilot until he plain passed out. Then I took control again and landed near the dock. We had quite a problem getting that limp body out of the airplane. Finally his friends pulled him out through the hatch, between the rudder pedals.

Recently I had received a letter from Manton with word that our mother had inoperable cancer. Since there was so little going on, and little prospect there ever would be, it didn't hurt my conscience to leave. I had been overseas more than a year, and Mother wasn't going to last very long.

On 5 October I went down to Bougainville to see Gen. Claude Larkin about going home. He was born and raised only about fifteen miles from Hubbard in a little place called New Era, so we had a lot in common. "Sheriff" was quite a colorful character, but well regarded in the Marine Corps, and his wife was equally notable. She was a welder in the Kaiser shipyards in Portland, and when the foreman was asked for his best worker to launch a Liberty ship, he chose Mrs. Larkin.

General Larkin gave me clearance for leave, and on 4 November I flew out of Manus in the Admiralties on a PB2Y, then on to Kwajalein and Johnson Island. Next day I arrived at Ewa near Honolulu and called Edna, who had been working again with the Powers agency in New York. She started toward Los Angeles, and a week later I was on a Martin Mars for the fourteen-hour flight to NAS Alameda.

I had left a car with relatives in Alameda and immediately drove to Los Angeles, where I spent several hours trying to find a room. Next day I met Edna at the Savoy. It was the first time we had seen each other in fifteen months.

People have often asked how our marriage survived so many long separations. I always said it was a matter of getting a good start: we had had six months together after we were married before I returned to combat. The other thing is our simple confidence and trust in each other. As you get older you become more tolerant of many things. But Edna's Irish luck certainly has stuck with her, and she's always had a halo around her as well.

Admiral Nimitz presents me with the Navy Cross at Guadalcanal on 1 October 1942. *U.S. Marine Corps*

From left to right, Maj. John L. Smith (CO, VMF-223), Maj. Robert L. Galer (CO, VMF-224), and myself after having received the Navy Cross from Admiral Nimitz. *U.S. Marine Corps*

Maj. John L. Smith, my CO in VMF-223 on Guadalcanal in 1942. *U.S. Marine Corps*

Grumman F4F Wildcats at an open-air repair shop on Guadalcanal in February 1943. The pilots' tents are in the background. *U.S. Marine Corps*

During my second tour to the Guadalcanal area, in late 1943, at Vella Lavella. I'm standing in front of an F4U-1 Corsair. *U.S. Marine Corps*

I set the world speed record on 25 August 1947 in this Douglas Skystreak (D-558-I) at Edwards Air Force Base. *Douglas Aircraft Co.*

In 1948 I was CO of VMF-122 (MCAS Cherry Point, N.C.), the first Marine Corps squadron to receive the FH-1 Phantom. *U.S. Marine Corps*

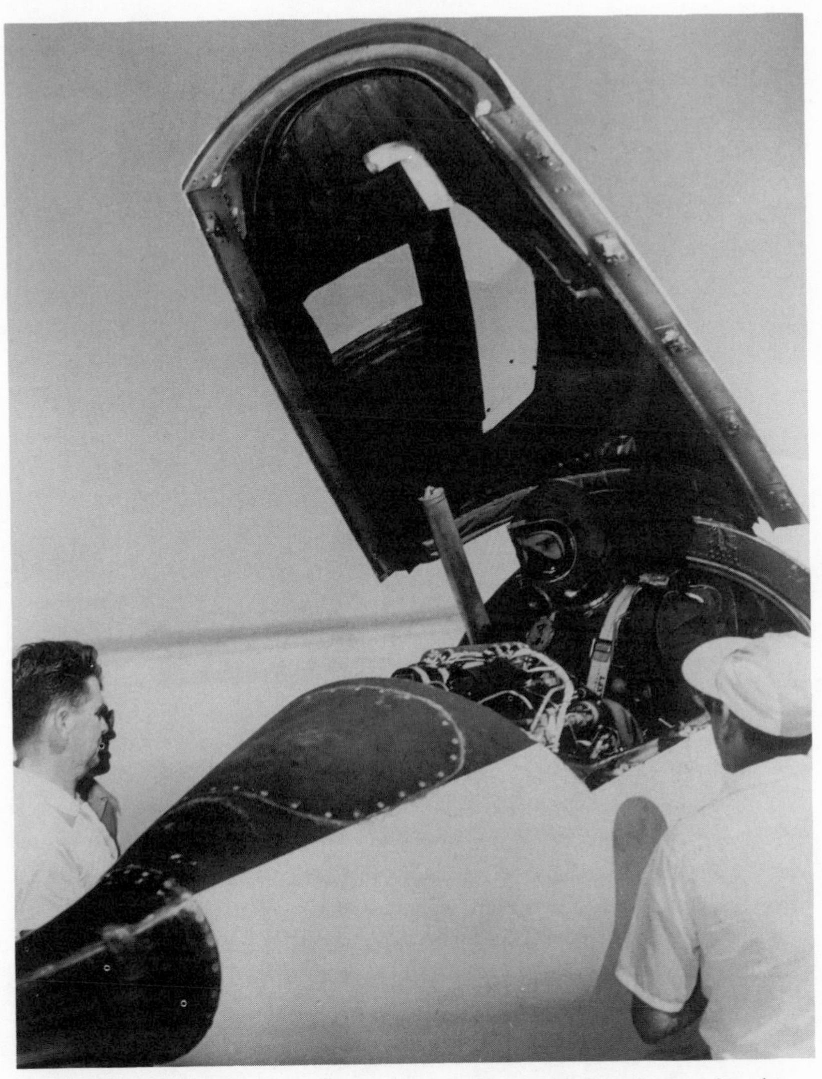

(*Photos left and right*) On 21 August 1953 I set the world altitude record in this Douglas Skyrocket (D-558-II) at Edwards Air Force Base. *Douglas Aircraft Co.*

Marine Corps aviator generals still got to do their own flying when I retired. Here, I have just returned to Cherry Point with a lieutenant colonel RIO after a cross-country to MCAS El Toro, California, on 15 August 1968. *U.S. Marine Corps*

My first day of retirement, 1 June 1973, celebrated with Edna and Commandant General Cushman. *U.S. Marine Corps*

I'm standing in front of an F4D Skyray at MCAS El Toro in October 1957. The F4D was my choice in unauthorized dogfights over southern California. *U.S. Marine Corps*

With Saburo Sakai in Tokyo in May 1971. We both flew in the Guadalcanal campaign, but Sakai was wounded and returned home shortly before I arrived in August 1942. *Tillman*

CHAPTER SIX

Pax River

In late November 1944 I received orders to an operational training job at NAS Jacksonville, Florida. There was enough time for Edna and me to visit my family in Oregon, so we drove north to see my mother. I was shocked at how poor her condition was. Even so, it was good to have about ten days with the whole family before Edna and I left for New York, where we arrived just before Christmas.

Before I set out for Jacksonville, I received a change of orders: I was to report to the new Naval Air Test Center at NAS Patuxent River, Maryland. The Navy had consolidated most of its experimental aviation projects at NATC—projects that previously had been handled at NAS Anacostia, D.C., and elsewhere. Considering my background in aeronautical engineering, the assignment made sense. But in those days hardly anyone had even heard of "Pax River." It had the advantage—and the bane—of being remote from just about everywhere.

I checked into NATC on 9 January 1945 while Edna temporarily remained with her mother in Brooklyn. I quickly felt at home with the senior aviators. Comdr. Paul Ramsey was coordinator of all flight test units and a veteran of early Pacific combat as skipper of VF-2, the Navy's famous "Flying Chiefs." Other marines included Lt. Cols. Des Canovan and Chick Quilter, both longtime friends. When I relieved Des he was reportedly the first marine to have survived a tour in Flight Test, and I intended to be the second!

I was assigned to the carrier section, where we were responsible for determining whether new aircraft were capable of operating safely from aircraft carriers. Other divisions included tactical test, armament test, radio test, and service test. The carrier section's chief was Comdr. Ed Owen, who was junior to me. Ed had led Fighting 5 off the second *Yorktown* (CV-10) and was an ace with eight kills. He asked if I objected to working for him, and I said not at all. As long as the head of any project knew what he was doing, neither I nor most pilots gave much thought to rank. However, Pax

River had its share of number counters who bristled at the thought of working for a junior, regardless of ability or experience.

Once Edna arrived from New York and we settled into our quarters, we began to explore the local environs. NATC was an isolated community. Outside the gate, at a wide spot in the road, was a place called Lexington Park, which had a few basic buildings—a diner and a small bank—and a population of about five hundred. The nearest "city"—about ten miles away—was Leonardtown, Maryland, the county seat, so social life was conducted at the air station. A temporary club was set up in the BOQ before the officers' club was built. Even getting to the officers' club took some commuting, as it was down on the point four miles or more from the main gate.

In February I flew the Bell YP-59A, an Army fighter and America's first jet aircraft. Almost anything on the base could outrun it, but still it was considered something special. I was about thirty-fourth on the list of people to check out in jets. Des Canavan and Al Hollar were the only marines ahead of me; Al was already at TacTest (tactical test) when I arrived at NATC.

My mother had taken a turn for the worse, so during March I gained emergency leave and caught an R4D transport to the West Coast. My sister met me in Portland and drove me to Woodburn Hospital where Manton and his wife were waiting. I remained in the area, knowing Mother's cancer was incurable, and she died on the fifteenth. I considered my mother a real pioneer type who was devoted to raising her children properly and seeing that they got the best education available. She was respected and loved by all who knew her.

After the funeral I caught a NATS flight out of Oakland and returned to Patuxent. It was good to get back to Flight Test; I felt that getting right back in the cockpit would be good therapy.

In mid-April all of us who were new to Flight Test attended lectures by Comdr. Sid Sherby, the chief engineer. He later started the Navy Test Pilot School and was its first director. Sid discussed aerodynamics and test procedures. We also had experts in other areas, including representatives from NACA. That same month I flew four new aircraft: a Howard GH-3, a prewar civilian type; the TBM-3 Avenger, a reliable old torpedo plane; the FM-2, last of the Wildcat line; and the F8F-1 Bearcat, which became my all-time favorite prop aircraft. Previously I had flown the ultimate Corsair, the XF2G-1 with a twenty-eight-cylinder Pratt and Whitney radial that was a fairly heavy engine for its power. It doesn't do much good to put a big engine in an airframe not designed for it; you get more climb but not much more speed than the standard model. I found the F8F was faster than the F2G and performed better.

An abundance of new aircraft arrived at Patuxent as the war progressed, and I couldn't seem to get around to all of them fast enough. I was assigned a project on the F7F Tigercat series, Grumman's twin-engine carrier fighter. I always liked the Tigercat as an airplane, but not particularly as a fighter. It was heavy and somewhat sluggish. It had power—two R-2800s—but was not very maneuverable, especially compared to the F8F.

At Flight Test we were involved in testing stability and control, fuel consumption, and everything associated with performance: measured takeoffs and landings, rate of climb, roll rate, responsiveness, and so on. A good fighter has almost neutral stability about all three axes. You can't have too much stability because that offsets maneuverability—the more stable an aircraft, the harder it is to maneuver radically.

Although we had a systematic approach to the job, our equipment was downright primitive. We had very little instrumentation in our project airplanes in 1945. Most of our information was gained with a stopwatch and a "strain gauge"—essentially a torque wrench—to measure stick forces, but the tools we had seemed to work well enough.

Experimental airplanes always have problems, and the Tigercat was no exception. During May I landed my F7F six times without hydraulics, which is to say without brakes. On 1 June my port prop governor went out and I made my seventh landing without brakes. Later that month I also flew the F7F-3N night fighter version for the first time.

On 15 June our class graduated from Test Pilot School, receiving diplomas and slide rules to mark the occasion. Ours was actually the first of five segments collectively designated "Class Zero," which ran through mid-1948 when the official Class One convened. Thus, Class "Oa" was the real pioneering segment of what is now perhaps the finest flight test school in the world. Back then our curriculum lasted only three months; today it is nearly a year.

A week after graduation I tore loose an elevator trim tab in an F7F. Six days later I landed single-engine after my port engine caught fire twice and got stuck in cruise pitch. It may sound dramatic today, but most test pilots quickly learned to take such things in stride. That was part of our purpose—to evaluate and solve problems *before* aircraft joined the fleet. I think some of us almost enjoyed handling emergencies—it was a break in the routine, and as close to combat as we could get.

One of the other satisfactions at NATC was meeting a variety of aviators. When I arrived we had a British Royal Navy liaison officer whom nobody particularly liked. But he was replaced by Lt. Mike Lithgow, who was an entirely different type of cat. Everybody liked Mike. He had a bit of a reckless streak, but he was an accomplished aviator. In fact, he became only the second pilot to make 2,000 carrier landings—a record that still stands.

NATC was only forty-five miles from Washington, so we got a lot of visiting dignitaries and were frequently asked to fly demonstrations. Mike liked to show off the Supermarine Seafire, and one of his favorite tricks was a slow roll on takeoff. One day he dished out of the roll and almost went into the trees. As senior, I was normally the flight leader who coordinated such airshows. I laid down the law to Mike—no more rolls on takeoff—and he knocked it off. Sad to say, a few years later, while chief test pilot for Hawker-Siddeley, he was killed in an auto accident. But he was popular at Patuxent and frequently participated in our socializing. We were a close-knit bunch, and since there wasn't much to do offbase, sometimes we'd play penny-ante poker after dinner. In the long run I think most of us broke even.

I got my first helicopter ride in July. NATC was evaluating a Sikorsky R-4B, which the Navy designated HNS-1. It was a little two-seater with about a 220-hp engine, and quite maneuverable. Charley Wood was the Navy pilot in charge of the helo section, and he made the mistake of asking me if I was interested in flying a helicopter. He gave me about three hours of dual and sent me solo—the only formal helo instruction I ever received.

I've been credited with being the first Marine Corps's helicopter pilot for more than forty years, and in fact I believed it a long time myself. But while comparing notes with Des Canovan I discovered that he soloed a helo during late 1944. If late really is better than never, I hope Des and the Marine Corps will take notice. Although I soloed in August 1945, I didn't get around to pushing the paperwork for designation as a helo pilot until July 1946. By then several other Marine Corps aviators had helo designations. But when I left NATC the first time I had 138 helicopter hours and was designated Marine Helicopter Pilot Number One because I had been the first to log the required 40 hours.

In 1946 I took over the helo section temporarily from Charley, pending arrival of his relief, Lt. Dave Gershowitz of the Coast Guard. Dave was a real pro with a helicopter, a young fellow who wasn't afraid to maneuver the machine. Flying with him was an education in helicopter agility, and years later I heard that he rated high on the Coast Guard's all-time list for life-saving rescues.

Another standout was Lt. Comdr. Don Runyon, a very good test pilot. I thought a great deal of Don, and went to New York that July to attend his wedding to a girl named Jean Wood. He had been a noncommmissioned aviator early in the war and, like many Naval Aviation pilots, was made an ensign in 1942. He became an ace in VF-6 and later had a tour in Hellcats.

Although Don's formal education was limited, his talent as a test pilot was second to none. He was also extremely modest. Following his death in 1986 I heard a story about a Grumman pilot who had flown with Don for ten years and had never even known Don had been in the Navy!

I can't help contrasting individuals like Don Runyon and Paul Ramsey, who knew the value of enlisted pilots, with some of the rank-conscious officers at NATC. A few of the latter had dug themselves foxholes in the States for the duration, either through personal influence or Bureau of Aeronautics connections. One particularly aggressive aviator from the Annapolis class of 1939 became so disgusted with the situation that he took off his Academy ring in 1942 and hasn't worn it since. It all adds up to the old saying, Why let rank lead when ability can do better?

Many of these considerations fell aside as the war ended in August. That month I took my flight physical for lieutenant colonel and, at age twenty-nine, learned my vision was 20/15. However, I could read a chart at 20/10 if the light was just right.

Other projects soon were upon me, leading to some of my most memorable NATC flying. On 22 August I took a Hellcat to Newark, New Jersey, to look over some German jets that had arrived—Messerschmitt 262 fighters and Arado 234 bombers. Four days later an Army pilot delivered a 262 to Patuxent and I was assigned the project.

My first 262 flight was 10 September, a forty-minute warmup for a ferry flight two days later as NATC received its second Messerschmitt. Later that month I went to Fort Meade, Maryland, to interview a couple of German POWs who had been sent over to help maintain the 262s. They arrived at Patuxent 5 October, under constant surveillance. I had to sign for them and, therefore, was responsible for their actions. But they were a cooperative crew who gave us no trouble. A third German, Carl Bauer, arrived the eighteenth, and I got to know him fairly well.

Carl had been Messerschmitt's chief test pilot in 1945. He had started work as the factory's junior test pilot early in the war when the company had only about ten pilots. Carl explained that he became senior by virtue of survival, as most of the others were killed trying to fly the Me-163 rocket interceptor. He stayed with us for only a few days, but the others remained until 6 November, when I presume they returned to Fort Meade and on to Germany.

On 7 November I was flying around by myself in an HNS-1 when I saw a 262 taxi out from TacTest. The pilot was Lt. Comdr. Bert Earnest, who had been the only survivor among Torpedo Squadron 8's six TBF pilots at Midway. The wind was down the short runway, 2-20, going right past the tower. I thought the Messerschmitt should use 13-31, which was probably

60 percent longer and wouldn't have suffered from the light wind. So I just sat back and watched the evolution, knowing it was going to be interesting.

The 262 needed roughly 160 mph for liftoff. (We had English instruments installed, as few of us knew metric.) Bert started his takeoff down Runway 20, which had a grove of trees at the end of a downslope on the far end. About two-thirds of the way down the runway the nose lifted off but flopped back down. Bert kept the nosewheel on the runway right to the end. The jet had slowed considerably, and when Bert popped back into the air it floated down over the slope into the trees. I paddled after him as fast as I could go in the HNS-1 and landed near the grove. I remember thinking, "What a way for poor old Bert to go, especially after Midway!"

I had just shut down and was ready to go into the trees for whatever was left of Commander Earnest when he started walking out, carrying his life jacket. He had hardly a scratch, but the only intact part of the 262 was the cockpit. The wings had been stripped off and the fuselage had gone snaking through the trees without hitting anything head-on. That pointed nose on the 262 had slid off stumps and tree trunks like a weasel through the woods. A guy with Bert's luck is destined to live forever.

We had lots of trouble with the 262s, mainly because of the Jumo engines and hydraulics. The engines required overhaul about every eight hours, owing not to design flaws but to critical material shortages. The project was canceled after my fifth emergency landing, but the design was well ahead of anything we had.

On 14 December I headed west to bring back a P-80 Shooting Star, the Army Air Force's first "real" jet fighter. I got three hops at Muroc Dry Lake (later Edwards Air Force Base) on the eighteenth and another the next day. On the twentieth I made a checkout flight from Lockheed Burbank to El Toro and back. Bad weather dogged me over the holidays, and I didn't get back to Pax River until New Year's Day 1946.

In February Col. Jack Cram arrived to relieve Chick Quilter at Flight Test. During the rest of my tour Jack was CO of the Marine Air Detachment and I was next in line, which suited me just fine—we always got along well. We began probing the P-80's strengths and limitations.

On 22 April I tangled with Comdr. Fred Bakutis, who flew an F8F. He was an experienced combat pilot, another ace and former squadron leader like so many NATC people, and he seemed confident that with the Bearcat's maneuverability he could force a standoff. But I soon found that by keeping up my speed in the P-80, I could pull up in a zoom climb. When he pulled up his nose to keep me in sight, pretty soon he had to push over or stall and I would be right on him. As always in aerial combat, headwork came first. But it was obvious that even the best recip fighters couldn't really compete with the jets. And the P-80 wasn't as sophisticated an air-frame as the older Messerschmitt 262. The main failing of the Messer-schmitt was material, not design.

In early summer, Ed Owen decided to leave the Navy for a job with in-dustry, so I became senior project officer for the carrier section. About this time, Lt. Comdr. Butch Davenport showed up. His given name was Merle, but he had been known as Butch since his days with Tom Blackburn's VF-17, the "Jolly Rogers." Except for Gus Widhelm's night-fighter squadron, VF-17 had been the first Corsair outfit in combat. Butch had been engineering officer and played a big role in making the F4U safe for carrier duty—an ironic twist because VF-17 was kicked off the *Bunker Hill* (CV-17) at the last minute and exiled to the Solomons where Butch shot down at least six Japanese planes. He was sent to the prestigious Empire Test Pilots School in Britain, along with Lt. Comdr. Jim Davidson, who had been number three in the carrier section. When Ed left, I moved up to number one and Jim took my spot.

Jim made the first U.S. carrier landing in a jet aircraft that July in a Mc-Donnell FD-1 Phantom. For obvious reasons, the Navy didn't want a ma-rine to be first aboard in a jet. Since the Navy lacked a suitable aircraft of its own, it was decided that tailhooks and affiliated gear would be added to Army Air Force types as an interim measure.

My main project was evaluating the P-80's carrier suitability. In August 1946 I began testing my P-80 in simulated carrier operations, and encoun-tered problems. On my first attempt I had an inoperable airspeed indicator and landed too hot, burning out the brakes and tires. Shortly thereafter I engaged the airfield arresting wire at fairly high speed, perhaps even at an angle, and broke the tailhook. But by the middle of September things were going well. I made ten arrested landings and we scheduled the shipboard trials for the end of October.

The new carrier *Franklin D. Roosevelt* (CVB-42), second of the *Midway* class, was available at Norfolk, Virginia, and I flew down there to have the P-80 loaded aboard. We put to sea the thirty-first and prepared for the tests next day, my thirty-first birthday.

I made four running takeoffs and two catapult shots in the Shooting Star, with five landing approaches. One was a planned waveoff at 10,000 pounds aircraft weight to see how the plane responded. Acceleration in early jets was notoriously poor, with sluggish "spool-up" time requiring delicate use of the throttle to avoid a compressor stall. But that wasn't the only concern. Because the hook was so short, I had to approach at an airspeed of 105 mph or less (the Army Air Force didn't use knots in those days). Stall was 98, so I shot for 102, intending to split the difference.

On one approach, turning up FDR's wake, I hit the burble and my airspeed dropped to 100 mph—just 2 mph above stall. But that was the only time it happened. The takeoffs and landings went pretty well and I returned to Patuxent. The Navy eventually obtained a batch of P-80s designated TO-1s and used them until FH-1 Phantoms and FJ-1 Furys became available.

Meanwhile I kept proficient in helicopters. In fact, at one airshow I made back-to-back flights in a P-80 and an HOS-1. The variety really appealed to me, and it was intriguing to envision the potential of such a versatile flying machine as the helo. I enjoyed giving a ride to Gen. Roy Geiger, my old boss at Guadalcanal, but my most notable flight of the period involved a ground officer.

During a demonstration at Quantico I said I was willing to use the hoist to lift someone into the cockpit. But I would be alone in the helo so I wanted a small, light individual. I got another lieutenant colonel, named Victor H. Krulak. He was about five-foot-six, and widely known as "Brute." It was not a sardonic nickname. He was all business, and to say that he was not universally adored would be an understatement. He stood under my whirring eggbeater and hooked onto the hoist, and I raised him into the cabin. Unfortunately, he had neglected to wear gloves, and I recall that the frayed cable caused him some distress. One of the spectators later suggested that I might have accidentally dropped my human cargo, but others held him in high regard.

One of those was Edna. She and Krulak got along particularly well, though as far as I was concerned it was a matter of two of a kind with a bit of ham in each of them. We saw a lot of Brute Krulak in later years, when the Marine Corps air-ground team had fully assimilated helicopters.

When Charley Wood left NATC I was the only helo pilot left in Flight Test. Consequently, I assumed all those duties in addition to my job as senior project officer in the carrier section.

Although the simple demonstration with Brute Krulak was little more than a stunt, it was representative of many NATC projects. Civilian leaders, the general public, and the military all stood to gain from exposure to new concepts and technology, and one of the best-publicized methods was revival of the national air races. At Cleveland in 1946, for instance, NATC pilots set a time-to-climb record for 10,000 feet. Butch Davenport and Bill Leonard, both top-notch aviators and fine gentlemen, set back-to-back records in a Bearcat. Bill was another of the long-term Pacific warriors, having flown F4Fs at Coral Sea, Midway, and Guadalcanal. At Cleveland he flew the mission the first day as required by the FAI representatives: full internal fuel and ammo. Brakes off to 10,000 feet was about ninety-six seconds. Next day was Butch's turn. He flew the takeoff and climb much as Bill had done, but benefited from a slightly warmer temperature. Consequently, he shaved two seconds off Bill's time. As far as I know, the F8F-1 still holds the piston-engine record for brakes off to 10,000 feet—barely a minute and a half.

But the emphasis increasingly was on jets. The new commanding officer of NATC was Capt. Frederick Trapnell, one of the Navy's early test pilots. He took a very personal interest in everything that went on, but particularly in Flight Test. In early 1947 I took an F7F to Muroc to evaluate North American's XFJ-1 Fury. I made six flights myself, and Trapnell also flew it because the Navy wanted a decision between the Fury and McDonnell's XF2H-1 Banshee. Trapnell and I agreed that the twin-engine F2H had more potential than the FJ, and McDonnell received a major contract based on our recommendation. In my opinion, this system for awarding contracts was far simpler and more efficient than today's multilayered bureaucracy. Only thirty-three FJ-1s were built, but they entered fleet service before the Banshee, mainly flying with Comdr. Pete Aurand's VF-51 at San Diego.

In July 1947 we began working to take away the world speed record from the U.S. Army Air Force. Col. Al Boyd, chief of Flight Test at Wright Field, was the current record holder with 623 mph in a P-80. Al was a fine individual, tremendously experienced and well regarded as a leader. I was probably chosen because I had more jet time than any other Navy or Marine Corps aviator, mainly due to the P-80 project.

Joining me in the record attempt was Comdr. Turner Caldwell from the Bureau of Aeronautics fighter desk. He had an excellent wartime record, first as an SBD pilot in the 1942 battles and later as commander of the Navy's first night carrier air group. However, I was concerned because Turner had no jet time.

I left Patuxent for El Toro in early August and went up to Muroc to look at the Douglas 558-1. The red-painted aircraft showed its heritage with an empennage reminiscent of the AD Skyraider. Ed Heinemann, the company's chief designer, had produced the Skystreak to exacting specifications. It was stressed for twelve Gs and, unlike most early experimental jets or rocket planes, took off under its own power. However, neither of the two -1s then available were ready so I headed up to Portland to see my family during the interim.

On 13 August Turner flew the P-80 for the first time and logged several other flights before trying the D-558. Our time at Muroc seemed wasted as far as we were concerned because we had arrived far too early. I was back at Muroc on the sixteenth and flew the second Skystreak. It was different, of course, but not especially difficult. Each of us made two flights next day and two more on the eighteenth. We even flew formation for a few minutes—perhaps the only time two D-558s ever did so.

On 19 August we were told the first record attempt would be the next day. The question remained open as to who would make the first run, so I said, "Turner, I'll give you your choice. You can make the first run tomorrow if you like, at whatever temperature you can get, and I'll make the second a few days later at whatever temperature I can get. Or, if you prefer, I'll make the run tomorrow and you take it later on."

Turner decided that he preferred to fly the next day. The ground temperature was 77 degrees when he made his four passes at an average speed of 640 mph. That beat the Air Force record by 17 mph, making it the first time in twenty-four years the U.S. Navy owned the world speed record. Turner then left for Los Angeles. I also left for a few days because it looked like cold weather was settling in.

A problem we were having with the D-558 was that we were not getting 100 percent rpm during the run. We got 100 percent on takeoff but not at maximum speed. Obviously there was some choking at the inlet, so I went to the tech rep to discuss the J35-C-3 engine and said, "How about setting the controls up a couple of percent? I'll guarantee to control the rpm at 100 percent on takeoff, using the throttle, and that'll give me enough during the runs. That would make a difference of about 2 or 3 mph; perhaps 5." The engine expert said he couldn't do that, but I thought I read him correctly. I shrugged and replied, "Well, that's OK, I can!" All it took was a little adjustment on the control. So he said all right—he didn't want me tampering with his engine!

On the twenty-fifth, when it was decided to make the run, the temperature at Muroc was ninety-four degrees—ten degrees below optimum. But we could not delay any longer, and ninety-four gave me enough advantage that I felt I could establish a new record, as the FAI required about 5 mph over the previous mark.

I took off at 1129 and made my four runs, two each way across the three-kilometer course. I had to stay below seventy-five meters while crossing the course, and I almost scraped the hilltops in the turns. Visibility was perhaps my biggest problem, as the canopy was almost pressed against my helmet and I couldn't turn my head very much. The ticklish flying was in the turns rather than the dash across the measured course. I noticed a little buffeting in the three-G turns, partly due to the airplane and partly due to rough air. However, the only turbulence was over the mountains—the aircraft was very stable over the lake bed. I landed eighteen minutes after takeoff, settling on at 160 or so and rolling for more than three miles. When I climbed out I was told by an elated Douglas-Navy crew that I had clocked 650.6 mph, or Mach 0.82.

There followed some publicity with the factory in Los Angeles, in Cleveland, and elsewhere. Naturally, I was happy with the title "fastest man alive," but that title changed frequently in those days. I was really pleased for Ed Heinemann and his dedicated, hard-working crew. In 1989 I visited Ed at his home in Rancho Santa Fe, California, and we discussed "might-have-beens." He said the D-558 Phase I was capable of supersonic flight—we could have been the first to exceed Mach 1, taking off and landing like a conventional airplane in contrast to the X-series, which required a mother ship and did fly supersonic later.

But despite sharing Edwards with the Air Force and Bell, we had no hint of the X-1 project. So, barely two months later, Air Force Capt. Chuck Yeager made history in the supersonic Bell, and my hat was off to him.

In mid-September I received orders to Cherry Point, North Carolina, to take command of the first Marine Corps jet squadron. Although I looked forward to returning to operational flying, I shared Turner Caldwell's assessment of flying the D-558 and on testing generally: "An exhilarating experience, and the nearest thing to combat flying since the war."

CHAPTER SEVEN

Jet Squadron

VMF-122 was to be the Marine Corps's first jet squadron, so naturally I was pleased to become the commanding officer. Edna and I made the 340-mile drive from Patuxent River to Cherry Point, North Carolina, and quickly settled in. Two days after we arrived—26 September 1947—I relieved Martin Ohlrich, who asked if he could stay on as my exec. I had not known Martin very well prior to this time. Previously he had been senior to me, but he had been passed over for promotion so I was now senior to him. It could have been a sensitive situation, but we got along well and I enjoyed having him work for me.

By this time I had about 3,600 hours' flight time, including quite a bit in jets, and had flown fifteen experimental aircraft that never entered production. VMF-122 didn't get any of the new McDonnell FH-1s until November, so we continued flying Corsairs. The Phantoms dribbled in by ones and twos, and we ferried some of them from the factory at St. Louis ourselves. Basically, we were disappointed in the FH-1 because of its poor performance. It was an easy plane to fly, but the two engines were sensitive to maximum power. I limited takeoffs to 96 percent RPM, which only gave us about 90 percent of available power but doubled the time between overhauls. The primary reason for limiting power was to increase engine life—a big factor in all early jets. Also, we eventually discovered the Phantom would take off single-engine on full internal fuel.

My former Pax River colleague, Bill Leonard, had the Navy's first FH-1 squadron at NAS Quonset Point, and he wrestled with the FH-1's problems, too. Thus, though the Navy and Marine Corps technically had joined the jet age, the available equipment was sorely lacking in combat capability. The FH-1 was no more than a technological stepping-stone, but we had to find some way of putting it to use.

Despite the FH-1's shortcomings, there were many pilots in the wing who wanted a slot in VMF-122. So I established a rule that, in order to qualify for the squadron and fly the Phantom, each pilot needed at least 1,500 hours' total time, with most of that in fighters. That took care of most of the aspiring jet jockies. I reviewed the records of the remainder and chose those I felt were most qualified.

I was fortunate that MAG-14 had enlightened leadership. When I took the squadron the group commander was Col. Jody Carlson—a rather temperamental person. However, he didn't interfere with the way I ran VMF-122. He was replaced in November by Col. Paul Putnam, who was a quiet type and a real gentleman. In all the time I worked for him I never once saw him lose his cool.

Although I always enjoyed the Corsair, I looked forward to becoming more proficient in the Phantom. I also retained my interest in helicopters. Cherry Point got a search-and-rescue helo about the time I arrived, and I looked up the major in charge of it. He said, "I know you're a helicopter pilot, but we just received a directive saying that only those with a helo designation can fly them. Do you have a designation?" I replied that I didn't know anything like that was required, as it was probably the first time the Marine Corps had issued such an edict. So I asked which authorities issued helo designations and was told Quantico, Lakehurst, and Patuxent River. So I sent a transcript of my logbook to Pax River and they forwarded it to the Bureau of Aeronautics. Next thing I knew I had my helicopter designation and a letter saying, "Our records indicate you are the first Marine to qualify for a designation as a helicopter pilot."

With the excitement and bustle of reorganizing the squadron, it was nice to have a pleasant living environment. Edna and I had good neighbors at Cherry Point; next door was Martin Ohlrich, and about three houses down was Bob Galer, who had been skipper of VMF-224 on Guadalcanal. As at Patuxent, there wasn't much outside the main gate so most social activities were held at the base. We were all good friends and thoroughly enjoyed each other's company.

Not all the fun of that tour directly involved flying. Shortly after arriving at Cherry Point I heard from Comdr. John Hyland, Patuxent's assistant director of Flight Test. He had two tickets to the fourth game of the World Series and suggested we attend, as the American League Yankees now led the National League Dodgers two games to one. We flew an SNB up to Floyd Bennett Field, which was practically within a stone's throw of Ebbetts Field. Edna had grown up in that area, so I was acquainted with it.

Johnny and I were treated to a terrific game. In the bottom of the ninth the Yanks led 2–0, though Brooklyn had runners on base. The Yankee pitcher, named Bevens, had allowed some walks but was just one out away from the first no-hitter in World Series history. The count was three and two on the Dodgers' third baseman, Cookie Lavagetto. Bevens uncorked a fastball and Lavagetto doubled two men home. When the game was over

Brooklyn had evened up the series, though the Yanks prevailed in seven games.

A couple of months later I learned the pitcher's full name—Floyd "Bill" Bevens from Salem, Oregon. I had played with him on the Hubbard High School team when I was third baseman! In those days nobody in the league wanted to bat against Bill because he had a sizzling fastball and not much control. Obviously he had improved a lot to pitch for the Yanks, who virtually owned baseball in those days. He later said he had been ordered to pitch Lavagetto that fastball, against his better judgment. I met Bill's son some time later and asked to be remembered to him, though I'm sure Bill preferred to forget that game.

On 1 March 1948 I was sworn in by Colonel Putnam as a lieutenant colonel—again. The first time I made light colonel—a temporary wartime promotion—was just before the end of the war. But then about 150 of us had reverted to major and had to wait for the selection process again. Although the promotion was personally satisfying, my main concern was still finding a legitimate use for the FH-1. Militarily it couldn't compete with the F4U-4, so we continued flying twelve Corsairs and twelve Phantoms.

At this time I began investigating the Phantom's aerobatic ability. I started with loops and rolls—even a loop to a landing—and then built two- and four-plane formations. Four of us made our first appearance in a formation flight for a change of command ceremony at Camp Lejeune. The wing commander, Major General Harris, saw our formation loop and called me to his office. I was a little concerned because we had no authority from anyone to do our aerobatics, but he said he liked our little routine and wanted us to continue. When I stated we had no authority for our team he said, "I'll get it," which he did.

The Phantom was a good airshow plane because of its slow-speed maneuverability. It could perform a routine well within the perimeter of most airports, and since I loved aerobatics this seemed a good way to sell the Marine Corps to the public while indulging myself. In early April I selected three other pilots and we began practicing for our first show at Floyd Bennett Field, New York, scheduled for 14 May. For a while it looked as if we'd have to cancel because the Navy restricted the Phantoms to level flight following a fatal crash, but the restriction was lifted almost immediately. We arrived at Floyd Bennett, made our airshow debut, and returned home without trouble. Between then and year end our little team performed at sixteen or so locations around the country.

During a cross-country in June I landed at Wright-Patterson Field near Dayton to visit Pat Fleming and his wife, Neville. Pat had been a lieutenant commander at Flight Test at Pax River, and I had gotten to know him well.

He was a very interesting person and I took an immediate liking to him. He had finished the war as the Navy's fourth-ranked fighter ace in a tie with Alex Vraciu—both had nineteen victories. Pat was ambitious in a nice sort of way and capable to boot, sometimes a rare combination. Regrettably, however, the Navy lost Pat to the new Air Force since interservice transfers were permitted. I believe that Gen. Curtis LeMay was personally involved in luring Pat to the Air Force, as Pat made lieutenant colonel shortly after transferring.

With his combat and test experience, Pat was a natural for Wright-Patterson, where he became chief of the fighter section under Col. Al Boyd. I frequently dropped in on the Flemings while transiting that area, and tried to keep in touch at other times as well. A few years later Pat transferred to SAC and made full colonel. Quite a change for a fighter pilot, but obviously he was going places in the Air Force.

Our aerobatic team was always having problems with our FH-1s, so we toured with an R5C transport that carried mechanics and a starter unit. However, on 28 July we had self-made problems in an airshow at NAS Oceana, Virginia. During a box roll Lt. Nelson Brown slid across my tail and we collided. I found that my only pitch control was on the left side, but I had to stretch the control wires with excessive stick force. On landing I had both feet behind the stick to keep adequate forward pressure—no small trick for a six-foot-two-inch pilot in the FH-1's cramped cockpit.

When I crawled out and examined my airplane I found a curl in the starboard stabilizer and elevator. We had to strike the aircraft because the fuselage was warped. Repairs would have required disassembling and rejigging the entire plane, so having McDonnell build a new one was cheaper. Brown's airplane wasn't damaged much—just a crease the length of one wing's undersurface. He flew back to Cherry Point, and we changed the wing there.

On 28 August we flew ten FH-1s to Cleveland for the air races, practicing for the event the six days beforehand. Although VMF-122 had no particular problems, there was one interesting incident. A loosely organized team called "The Flying Admirals" was composed of three senior aviators—Admirals Soucek, Gallery, and Cruise—all of whom flew Phantoms. That day it was pretty hazy and visibility was limited. Consequently, they came across the opposite end of the field just as I took my team across, and we raced for each other head-on. Actually, it wasn't a problem as long as the admirals held their altitude, so I led my team down on the deck and squeezed in between the three Navy jets and the runway. It must have looked like split-second timing from the grandstand, but to this day I'm not sure if the spectators realized they had almost witnessed a potentially spectacular midair collision.

My time in VMF-122 soon ran out. In late September Doc Everton arrived as my new exec, and just after Christmas I was transferred to MAG-14 as operations officer. Col. Ed Pugh was the CO and Bob Galer had become his exec, so I was with old friends. Not much happened for the next year. My main program was getting all MAG-14 pilots instrument-qualified, including Colonel Pugh, which proved quite an undertaking. But he got his white card since I wasn't about to quibble over giving him an "up"!

As time allowed I slipped over to -122 for some solo aerobatics and even cadged some hops from Johnny Hyland, who now was assistant director of Flight Test at Pax River. He allowed me to keep current in helos and check out in newer tactical aircraft.

In June 1949 I was ordered to England as the U.S. Marine Corps representative at a fighter conference at West Rayham. I teamed up with Gus Widhelm, whom I'd known at Pax River. Gus gave new meaning to the adjective *colorful.* He was one of the best aviators the U.S. Navy ever produced, and if he liked you he would move heaven and earth to help, and regulations be damned. Butch Davenport once casually mentioned that he was having difficulty moving his family's belongings from Patuxent. Overhearing Butch's comment, Gus chipped in, "I can get a B-29. Would that help?"

I couldn't believe how bitter cold Britain was in June—I thought I might freeze to death before the conference ended. But there were amenities. I was assigned a "batman," which I thought was pretty unique. When you woke in the morning there would be a cup of hot tea by your bed. The Brits really took good care of us. And how they partied! Cocktail parties typically lasted until 0400, though I seldom remained past 2300. On the last day of the conference the socializing began at 1700 and lasted for eleven hours!

Everybody left after the conference, but I had made arrangements to fly the Sea Vampire and Meteor IV in order to compare British jets with ours. While in London the next day I got a cable from Edna announcing the birth of our first child, Lyanne.

The day after I became a father I met Comdr. Eric Brown at the Empire Test Pilot's School, who was working on the rubber deck project. Eric, whose nickname was "Winkle," had a wealth of operational and test experience. He had flown Wildcats (Martlets to the British) from the Royal Navy's first escort carrier and was a contemporary of Mike Lithgow, the exchange pilot at NATC. Eric still holds the world's record with some 2,400 carrier arrested landings, so he and Mike were the two top "trappers" and probably will remain so. Eric's wife was a vivacious Irish girl named Lynn. She was just as extroverted as Edna, while Eric and I were both much more reserved. Anyway, with help from Eric and from Bee Weems, my assistant in the carrier section of NATC, I got to fly a glider and take a hop in a DeHavilland Mosquito. Then it was back to Patuxent on a C-54, counting the stops until midnight next day. When I returned to Cherry Point on the twenty-fifth Edna was still in the hospital, and that was the first time I saw our new baby. I took them both home next day.

The rest of the summer was busy with the growing family and new airplanes. In August I took an F4U to Wright-Patterson to visit Pat Fleming. He arranged for me to check out in the F-84E, which I took to 42,000 feet and Mach 0.81. Then I hopped into an F-86A, and reached 47,000 feet and Mach 0.94. It probably never occurred to me at the time that those were going to become "the good old days." Swapping rides with foreign and U.S. Air Force pilots was one of the joys in the community of flying, and we didn't expect that regulations would tighten up in a few years.

Another joy was giving Edna a helicopter ride. Previously I had taken her on a flight in an F7F, though the view from the rear seat wasn't terrific. In those days military aviators could take their wives for flights annually, although tactical aircraft weren't supposed to be flown and each hop was to be limited to half an hour. As I recall, when Edna had the Tigercat ride she was about four months pregnant. But that didn't keep me from putting the airplane on its back to show her how it felt.

In September 1949 I took an F9F-2 Panther to the West Coast for a week of promoting Navy and Marine Corps aviation. I made a dozen or more speeches in that time—not exactly my idea of a fun trip. But it involved two transits of the continent in a jet fighter, so I figured I came out even.

The first leg of the return trip upped the ante, however. At Portland Air Base my Panther developed a dead battery, so I had to talk the National Guard out of a new one. The dead battery should have made me suspicious. The problem delayed my takeoff until dusk 1 October, filing for NAS North Island, San Diego. Near Los Angeles my electric compass started spinning, my radio went dead and all my lights went out. I'd had an electrical failure. That was the bad news. On the other hand, I knew I had plenty of fuel—mostly in my wing tanks. That was the good news. But without instruments I couldn't tell if I was drawing from them. However, I soon found out because my engine quit over Oceanside, fifty miles north of San Diego.

At 27,000 feet I had to make a quick decision: turn back for MCAS El Toro or proceed to North Island. Since I had filed for the latter, I continued on and set up my best glide speed. It worked well, as I had more than 5,000 feet overhead the air station. I decided to buzz the tower to let them know I was coming in because I had no power, lights, or communication.

I pulled up from my pass at the tower and wrapped the Panther into a 270-degree turn. Everything was all right until base leg when the windscreen froze over. I had to skid the plane so I could look out one side, but I made a successful deadstick landing.

I was a mile down the runway when I stopped, and I waited for somebody to come out and give me a hand. Nothing happened. Finally I jumped out and walked a mile to the tower, leaving my airplane on the runway. I walked in and asked the duty officer, "Do you realize you've got an airplane on that runway with no lights? Anything trying to take off or land is liable to run into it." That woke him up. I added, "By the way, I made a pass at the tower to alert you. You had a flight plan on me, didn't you?" The lieutenant said, "Oh, yes." So I asked him who he thought buzzed the tower and he replied, "Oh, I just thought that was somebody flathatting."

There was an F9F squadron at North Island, and they got busy on my plane next morning. They couldn't find anything wrong with the electrical system; just a dead battery. Everything else checked out, so I took off and continued east, refueling at Denver with an RON at Glenview. But the excitement wasn't over.

Two days later I continued to Cherry Point, filing IFR on top. Everything was fine until the magnetic compass started spinning, indicating another failure. Immediately I contacted Raleigh Flight Following and canceled my IFR flight plan. I had no sooner done that than everything went dead. Now I was on top and had to find some way of getting down through the cloud deck without instruments.

If the overcast is very thick, a descent without instruments inevitably leads to a graveyard spiral. I knew I had a 2,500-foot ceiling underneath and was contemplating the old airmail pilot's trick of spinning down through the overcast, but fortunately I didn't have to do that. I found a hole and got down safely.

I still wanted to find the cause of the electrical problem, so the next day I had the crew tie the airplane down and turn up normal power. After forty-five minutes the electrical system went dead again. The mechanics found that the reverse current relay was causing the battery to overcharge and short out. Pulling off the night landing at North Island without power, instruments, or lights was the sort of unexpected challenge that kept flying so attractive. But I definitely didn't want to do it again.

About this time I got a call from Captain Trapnell, commanding officer at NATC. He asked if I would like to return to my old Flight Test job at Pax River. The prospect had tremendous personal appeal to me, but I was concerned that returning to the same job wouldn't look good on my record. However, once I made my decision I didn't have to do anything—Trapnell made all the arrangements. After the customary military ritual of frantic packing, address changes, and good-byes, Edna took Lyanne to Brooklyn until I got settled.

I reported to Patuxent River in December 1949 and quickly got back into the saddle. Next month I had a couple of flights in the F6U-1 Pirate, Vought's first jet. It was considerably underpowered, and the program was canceled after just thirty-three aircraft had been delivered. But the company's next design occupied a lot of our time.

In February 1950 I was joined by Lt. Comdr. Joe Rees in the carrier section. Joe had hunted submarines in the Atlantic and once had to swim away from the escort carrier *Block Island* (CVE-21) when she was torpedoed by a U-boat. We flew to Ardmore, Oklahoma, to check out the F7U Cutlass, a radical twin-tailed fighter that Vought hoped would be the jet equivalent of the F4U. I flew the Cutlass four times and Joe flew it three, and the more we flew it the less we liked it. Consequently, we returned to NATC and recommended that the airplane not be purchased. That started quite a feud.

Comdr. Noel Gayler of the Bureau of Aeronautics had flown the F7U a couple of times and was enthusiastic about it. On paper, it looked great: twin afterburning engines, supersonic speed, and exotic good looks. Its performance came from the afterburners—the first on any Navy production aircraft—but endurance was limited to about thirty minutes in burner. Additionally, the F7U's flight-control system caused problems, and its range and carrier-landing qualities left a lot to be desired. Nevertheless, Noel, who had no flight-test experience, recommended that the Cutlass be bought, and Vought got a contract.

I could see the airplane was going to be a dog, so I went to Joe Renner at Marine Corps Headquarters and recommended he go on record as stating the F7U would be of no use to the Marine Corps. In those days, when a Navy airplane didn't work it was likely to be panned off onto the Marine Corps, and I didn't want that to happen this time. By then Joe Renner was head of military requirements, and he declared there was no use for an airplane that lacked external stores and which couldn't be used in the attack role. This position was approved by the Commandant of the Marine Corps.

The upshot was that the F7U only made a half-dozen carrier deployments in its three-year fleet service. Of the 330 or so jets produced, only one logged more than 500 hours. In that time the Cutlass killed a lot of good young aviators and became known as the "Ensign Eliminator." The tragedy was that the losses were so unnecessary.

I began working on a variety of intriguing projects, including Douglas's twin-jet night fighter, the XF3D-1 Skyknight, and the variable-stability

F6F-3 Hellcat, which could be programmed to represent the expected handling qualities of new designs. The Hellcat's stability characteristics could be adjusted both laterally and longitudinally, something now routine in flight-test circles.

There were "good deal" jobs, too. In late May four of us comprising a U.S. delegation flew to Stockholm for the FAI conference. Rear Admiral Lonnquist and I were the naval reps, and we were joined by Brig. Gen. Al Boyd of the Air Force and Jackie Cochran. She was about the pushiest, brassiest woman I had ever met. Her attitude was that if you weren't Air Force you weren't worth considering—an attitude Al Boyd definitely did not share. Perhaps the most intriguing aspect of the trip was a tour of Saab Aircraft's underground maintenance plant, but I also spent a fascinating day and night on board the HMS *Illustrious* seeing how the British operated.

In September we put on an airshow for visiting VIPs. I was flying a Bearcat we called "Old 049," the same one in which Butch Davenport had set the time-to-climb record of ninety-four seconds to 10,000 feet. This show was the first time I looped an F8F off the catapult, landing beside the cat. It was probably the world's shortest airshow performance, but it was a real attention-getter.

We also got two F-86As from North American as pace aircraft, and I ferried one to Pax from Van Nuys. On 5 October, while at altitude, I made a few Mach 1.05 dives, which caused some confusion on the ground at Patuxent. It was the first time anyone at the base had heard a sonic boom, and NATC blamed it on Dahlgren, everything and everybody but me. Finally I had to confess that I had been doing some supersonic dives out over Chesapeake Bay.

But along with the fun and games was the lurking specter of tragedy. My first Pax River tour had been relatively free of losses, but the second stint brought a rash of fatalities. We lost a pilot from the Bureau of Aeronautics in the Douglas XA2D-1 turboprop in late 1950, and then in January 1951 my assistant, Bee Weems, was killed in a water skiplane accident. Bee was a good aviator and a top-notch officer, and I had been pleased to have him when Joe Rees rotated out. Bee's father had devised the Weems system of navigation and lived near Annapolis.

In March two pilots were killed in a Martin P4M patrol bomber, and in April we lost a pilot when his AD-2 Skyraider unaccountably dived into the bay. We never learned the cause, though we suspected the rubber drop tanks under the fuselage may have been a factor.

Later we almost lost Lt. Comdr. George Duncan, chief engineer of my section. While landing an F9F-5 aboard the USS *Midway* (CVB-41) he had a spectacular rampstrike and his aircraft exploded in a huge fireball. The

Panther was torn apart and the nose section rolled up the deck. Everyone who saw it figured that was the end of George, but when the fire fighters got to the cockpit they were astonished to find him very much alive, though burned. The spectacular footage from the crash has appeared in at least two motion pictures, including *Midway* and *The Hunt for Red October.*

However, we learned to take the good with the bad; in late March Edna gave birth to our son, Bruce.

NATC retained cordial relations with the British, and eventually Edna and I renewed acquaintances with Eric and Lynn Brown when they arrived on exchange duty. Eric probably was the most experienced test pilot flying at that time, and his notoriety only increased with time. He was a dapper Scot who spoke with a marked Edinburgh accent, but his speech and choice of words were as precise as his flying. Despite the professionalism that characterized NATC, there was always an undercurrent of competition which, within limits, was healthy. Winkle certainly held his own, representing the Royal Navy in fine style. Edna and I quickly formed a close relationship with the Browns, one that lasted well beyond Pax River.

In December 1951 I passed 6,000 hours' total time.

CHAPTER EIGHT

Flight Test—Again

By early 1952 I was occupied with two very different projects at NATC. The first—and perhaps the Navy's greatest priority—was getting nuclear-capable aircraft ready for carrier operations, and that meant the AJ-1 Savage. The big North American bomber was a curious animal with two reciprocating engines and one jet. But it was the only game in town for getting atomic weapons into carrier aircraft.

In February I went aboard the carrier *Wasp* (CV-18) for some launches and arrested landings, testing various control configurations. With the control boost off, mainly using trim tabs, I wrestled the Savage around the carrier pattern—partly by brute strength. One landing in particular was a controlled crash, but the airplane took it without protest. The AJ-1 weights were 47,000 to 49,000 pounds for these tests—by far the heaviest for routine carrier operations up to that time.

The second NATC project was Grumman's AF Guardian, the first dedicated, carrier-based antisubmarine aircraft. It was a big prop-driven design typical of the famed "Iron Works," and had entered fleet service less than two years earlier. The Guardian came in two versions, the radar-equipped AF-2W "hunter" and the AF-2S "killer," and I was conducting spin tests on a -2S loaned from the factory.

Our test aircraft had the underwing radar and searchlight removed for the spin tests, which it performed without difficulty. But I thought it only practical to finish the series of tests with the external stores in place, as they were employed in squadron service. For reasons I couldn't explain—perhaps a premonition—I had a pararaft fitted in my seatpack parachute. This was the first time I had taken that precaution in six months or so.

The final spin tests were to be flown 1 April at about 12,000 feet over Chesapeake Bay, with Eric Brown and Lt. Comdr. Bob Clark chasing me in an SNB-5. The Guardian had a fairly abrupt stall but no serious post-stall problems. Previously I had kept the spins to one and a half or two turns, but this time I allowed the AF to progress to two and a half turns to the right. Recovery was effected with no problem in a half turn. I checked the radar and searchlight housings, determined they were intact, and prepared for the left-hand spins.

I made two efforts at the second set of spins, but they wouldn't develop fully and I only generated spirals rather than actual spins. So I tried a departure stall from a climbing left-hand turn at about 11,000 feet. This time the big Grumman fell off to starboard. I let it go two turns before applying corrective action, and on the second rotation the nose came up. I was in a flat spin and knew I had real trouble.

I tried the standard remedies. Neutral controls, full power, stick forward, everything. No good. I rode it through several turns—I didn't know exactly how many—but I did know one thing: I had to get out of that airplane.

The AF-2S had side-by-side seating and I was in the left seat. Unfortunately, the centrifugal force of the spin was so great that I couldn't get out that side. I had to crawl across the empty seat and try the far side. But the force of the spin made this take too much time. The water was rushing up to meet me as the dizzying spin held me inside. I was trapped.

From their vantage point in the "bugsmasher," Eric and Bob counted ten to twelve turns as the Guardian stubbornly maintained its nose-level attitude. They watched it all the way down until it hit the water. There was no sign of bailout. They circled the crash site and called for a rescue aircraft.

I pulled myself across the cockpit, into the twirling slipsteam across the starboard wing. Somehow I avoided snagging my chute on anything, and as soon as I reached the wing I pulled my D-ring. There wasn't one second to spare.

My parachute blossomed and immediately I hit the water. In fact, I didn't even swing once. I went in right beside the aircraft, so neither Eric nor Bob saw me splash down. They did see my canopy streaming under water, however, and that gave them some hope. Then they spotted me.

Had I been over land I would have broken both legs at the very least. Luckily, I wasn't seriously hurt, but I was in danger of drowning or freezing. I had trouble getting out of my chute because a breeze was blowing, and I certainly had no time to unsnap my leg straps before hitting the water. However, this was my second bailout, so I knew the procedure. With a water temperature of 46 degrees, Chesapeake Bay was far less hospitable than Ironbottom Sound off Guadalcanal had been ten years before.

If it wasn't for the raft in my seatpack I probably wouldn't have made it because I swallowed quite a bit of water and was suffering from the cold. I barely managed to inflate the raft and haul myself aboard. All I could do was lie there, trying to catch my breath while my teeth chattered away.

A few minutes later I heard an airplane. Eric and Bob had called the base when they saw the crash, and, as luck would have it, a seaplane was landing at that moment. Bert Stahl of the non-carrier section of Flight Test had on full power and rpm, driving right for me. A very happy sight to a guy who was almost frozen.

Bert pulled me aboard and delivered me to the hospital. I had no serious injuries, but I was still in mild shock. It took a while for me to thaw out.

We were still involved with the F7U program, and I made frequent trips to Vought Aircraft Co. in Dallas. On one flight in June I flew a Cutlass the day before Bert Henderson was killed in one. I made two more flights that same day, more convinced than ever that the F7U just wasn't going to work. I discussed it with Paul Thayer, formerly Vought's chief test pilot, and then flew my F-86 back to Patuxent.

Then on 23 August the gremlins popped up again. I took off in an F9F-5 to chase Douglas pilot Bob Rahn on an F3D-1 demonstration flight. I noticed engine vibration and shut down, suspecting I had lost a turbine blade (it turned out to be a faulty fuel control). I set up for a deadstick landing, but I had flap problems and couldn't quite reach the runway. I hit hard and, basically, broke my back—a compression fracture of the number three lumbar vertebra. The Panther was a strike.

I was placed in a body cast from hips to shoulders, and about a week later I was put in an even tighter cast for eight more weeks. It was removed 27 October, but my medical status was restricted duty, meaning I was no good for Flight Test. I was faced with the very real prospect of no longer flying, which for me would mean leaving the Marine Corps. I gloomily tallied my working life of 6,500 hours, including 850 in jets, 200 in helicopters, and 450 on instruments.

Rear Adm. Mel Pride was head of NATC, and he agreed with Marine Corps Headquarters that I should be transferred to MCAS Quantico. Meanwhile, I made my annual safari to Maine to go deer hunting with Edna and Jane and Sid Sherby. We all had a good time together, and secretly I was pleased Edna took to the outdoors that way—especially for a New York girl. Usually I hunted with Flight Test guys like Don Runyon and Whitey Feightner, or hunted alone, so in Maine I managed to sneak off by myself a few times to do some serious hunting.

We moved from Patuxent in late November and I became operations officer at Quantico's Brown Field, a purgatory I endured for the next nine months. I didn't mention my restricted status to anyone, and if the authorities were aware of it they didn't care, because I flew 100 hours my first month there. It was a surprisingly loose arrangement. I often flew down to Dallas to visit Paul Thayer, usually in an F7F. We may have been in the jet age, but the Tigercat was still a terrific cross-country airplane.

By early 1953 the Navy realized that, even though I was no longer at Flight Test, I was the only pilot who could fly the new Douglas D-558 II experimental aircraft. A special full-pressure suit was required, and I was the only one who could wear it.

The Air Force at that time used a partial-pressure suit with a capstan arrangement that pulled the suit tight around your arms, legs, and torso, but which didn't pressurize your head, hands, or feet. The Navy opted for a full-pressure system. I had been at NATC in late 1951 when the D-558 II program got started, and, as one of the project pilots, I had gone to Philadelphia to be measured for the special full-pressure suit manufactured by the Clark Company of Wooster, Massachusetts.

In March 1953 my restricted status had expired, since it only applied to the first six months after I broke my back. The D-558 II tests weren't scheduled until the summer, so I was still in contention. Besides, Rear Adm. Apollo Soucek was the chief of the Bureau of Aeronautics, and a booster of mine. He'd been a pioneer test pilot himself in the 1930s. With the help of Admiral Soucek and my boss at Quantico, Col. Art Binney, I managed to stay current on the Skyrocket program. As operations officer at Brown Field I had access to some F7Fs and could fly out to Edwards AFB fairly often.

In early June I called Dave Clark, the owner of the Clark Company, to ask if he could speed up work on the pressure suit. A few days later I took a Tigercat to Wooster to pick up the suit and then headed west for a brief hiatus in Oregon before heading back to Edwards. As usual, I stayed overnight with my friend Carl Stiefel in The Dalles. The next night I got a call from my brother, who said the barn had burned down at the farm. I visited him en route to California, and while I don't have any way of determining such things, I can't imagine many test pilots' itineraries involved inspection of recently burned farm buildings.

On 15 June I went to the National Advisory Committee on Aeronautics (NACA) office to check plumbing on the pressure suit, but I learned that the D-558 II wouldn't be ready for another couple of months. NACA's project pilot was Scott Crossfield, who checked me out on the Skyrocket. In my estimation he was not only a very fine pilot but a talented engineer, and I was pleased to be working with him. Tests on the suit continued next month in Philadelphia, where I spent a lot of time at the Air Materiel Laboratory (AMEL) in compression checks and the like. We tested the suit in the low-pressure chamber at simulated altitudes of up to 70,000 feet, and I also tolerated a two-hour session in a Link Trainer with the suit pressurized. The latter session made no sense to me, as the actual flight time for the altitude record attempt would only be fifteen minutes.

Back at Edwards on 27 July, I prepared for the first D-558 II flight next day. I was dropped from a B-29 at 33,000 feet using jet power only, and two days later I used both jet and rocket power at 20,000 feet. On the thirty-first I returned to Quantico and was back at Edwards 5 August. But the Skyrocket was out of commission after Scott Crossfield's flight the day before, so I climbed back in my F7 and flew home. On the eleventh I went up to Philadelphia, picked up an AMEL engineer, and landed at Edwards at 1530 that afternoon after 10.3 hours. At least I was getting in a lot of good cross-country flying.

But the gremlins kept throwing glitches at the D-558 II. My flight scheduled for the twelfth was canceled because of overcast weather and next day the B-29 aborted with an oxygen leak. The latter could have been a close call for me. I was in the cockpit as the countdown started, and Scott Crossfield was inside the B-29. The count started at ten, but at eight I realized I had no oxygen. The mother ship's system had been cut off and the D-558's system hadn't taken over as planned. The copilot kept the microphone button depressed as he made the countdown, so I couldn't cut in. Fortunately, a crewman was on the deck watching me, and he saw my hand signal. He was able to get the pilot's attention and the countdown was stopped at three—maybe even at two. Another couple of seconds and I would have been dropped at 33,000 feet. I had gone to 27,000 feet in the chamber without an oxygen mask, but I don't know what I would have done a mile above that.

We jettisoned the fuel and landed with the Skystreak attached to the B-29. They discovered that the link between the pilot and the ship's system had come loose. It was simple to fix and we got going again next day.

At 1300 the B-29 took off, and this time it dropped me. But after engine ignition I climbed too steeply and only attained 72,000 feet. It would be four days before we were ready to launch again, so I ran up to Portland and back. It gave me time to think things over. The flight profile had to be exactly right; either too steep or too shallow and I wouldn't make it. Bill Bridgeman's record mission had been flown just right, but he'd had several flights in the aircraft. My experience was limited, and I wasn't counting on more than three flights.

I wanted to break Bill's record by at least 5 percent to have a significant increase. His altitude was 74,000 feet, so I needed 3,700 feet more than that. But I figured I wanted 80,000 feet just to be safe. However, on my second flight I climbed too steeply again and only managed about 75,000 feet. I had one chance left.

By now I had figured out what I had to do. There was no way to reach my goal without taking a little additional risk. Bridgeman had waited three minutes until he ran out of fuel, then shoved over the top to be well above stall speed. I decided to hold my attitude, thereby establishing a zero-G trajectory, because if there is no lift on the wings they won't stall.

The drop was routine. The engine lit off, I established my climb satisfactorily and flew a good schedule. I figured I went over the top at about 80 mph instead of the 140 or so indicated—quite a bit slower than during Bill's flight. I knew I was flying a good schedule because at burnout I was significantly higher than on the two previous flights. Instead of shoving the nose over, I held the same attitude until my speed had dropped off. However, I experienced zero gravity at the apogee so I didn't care too much about airspeed at that point.

The flight was over in twelve minutes, and upon landing I learned I had made 83,235 feet, which was as much as I could have possibly hoped for. It was an unofficial world's altitude record. But I didn't hang around. I jumped in my pet Tigercat that afternoon and headed for Quantico, arriving next morning after a fuel stop at Dayton.

However, Edwards beckoned again. At month's end I was back there, trying to break Mach 2. I had taken the Skystreak higher than anybody else had and now I was trying to go faster, too. On my first launch, 31 August, I got into violent oscillation during the pushover at fairly high speed. It was so rough that it unported the fuel, causing the rocket cylinders to flame out. I cut back immediately, slowed as quickly as I could, and aborted the flight. That was my first experience with such violent oscillation and I took it as a lesson learned.

On 2 September I was launched again from the B-29. I went to 70,000 feet and 1,140 mph, but that was considerably short of Mach 2. Next day I went to Wright-Patterson to fly an FJ-2 Fury in the three-day airshow, then returned to Quantico. However, I continued pressing for a few more flights in the D-558 II. I determined that I had been going a little too high before I pushed over, and that meant I was running out of fuel just a bit too soon. I went to the Bureau of Aeronautics to request one more chance, but they had run out of funding—each flight cost about $50,000. Navy research money had dried up, but a short time later Scott Crossfield broke Mach 2 in the D-558 II.

Thus ended my time in the world of experimental and research flying. It had been fascinating work, rich in opportunity and satisfaction at "pushing the envelope" of our knowledge in high-speed and high-altitude flight. The Korean armistice had been signed in July, and my disappointment at missing additional combat was eased with the challenge of test flying. I now had 7,000 hours, including 3,000 in Flight Test. But it was time to return to earth.

In late September 1953 we moved into our new home at Triangle, just outside Quantico. We lived there off and on for the next thirty years. I started Senior School that month, but it was pretty dull for me. Most of it dealt with Marine Corps ground operations, which was not very far up on my list of interests. However, one of the bright spots was a Navy doctor named Bill New. He had the best memory in the class and spent less time studying than anyone; he just listened to what the lecturers said. We really hit it off, and we've stayed in touch ever since. In a way, he provided a lesson far removed from the curriculum, for even in nonflying duty there was benefit to be gained if I just took time to look and listen.

CHAPTER NINE

Banshees over China

While at Headquarters in February 1954 I talked to Gen. Oscar Brice, the director of Marine Corps Aviation, and made a pitch to attend the USAF Gunnery School at Nellis AFB in Las Vegas, Nevada. John Condon had been the previous Marine Corps candidate for the slot and recommended that in the future nobody over thirty-five be sent. I was thirty-nine at the time and had over 7,000 hours, which was more than Condon had.

Eventually I got the nod; it was decided to send me to Nellis before I proceeded to duty in Korea. Senior School finished in early June and I headed west with a friend from Chicago, John Watzek. The highlight of that trip was the T-Cross Ranch at Dubois, Wyoming, where we had a wonderful time camping, fishing, and horseback riding. In the episode I remember best, I was walking along a creek with my fishing rod when I heard a shrill sound and looked up. Coming down the slope was a calf elk with two coyotes in hot pursuit. The calf came right past me. I pulled out a small pistol and started shooting at the coyotes. I didn't hit them, of course, but the noise scared them off. John said the calf ran right between him and the tent in our camp and disappeared into the brush. I have read that coyotes don't live off elk or deer, and while that's the only time I've seen them chase an elk, they certainly will run a deer to exhaustion and then kill it.

I arrived at Nellis AFB on 30 July 1954 and reported to Lt. Col. John Girardo of the USAF Gunnery School. He looked at my logbook to see what sort of experience I had and responded favorably. I had 7,300 hours' total time, with almost 1,000 in jets, which I inferred was as much as anybody on the base. John and I became good friends; he was a great guy who eventually retired as a lieutenant general.

There was no problem at the school, flying F-86s, but the Air Force maintenance system irked me. You could walk out to your airplane and ask the crew chief about its condition, but usually he knew nothing about that particular airplane. The Air Force had centralized maintenance, which was all right for bombers or transports but created havoc in tactical units. In the

Marine Corps the plane captain stayed with his own aircraft for months or even years and knew it inside and out.

One morning at Nellis I made walkaround inspections of three planes and never did fly; after I turned down those three none remained available. John Girardo shared my concern but there was nothing he could do about it.

Flying two or even three sorties per day quickly honed my rusty gunnery skills. In the 20,000- to 25,000-foot patterns I began with a mere 5 percent hits but peaked at 35 percent before moving on to bombing and strafing. For someone who loved shooting as much as flying, Nellis was pure fun.

I made two cross-countries in August and helped Edna and our children pack for the trip to Nevada. Shortly after we got settled, two Air Force colonels moved into the guest quarters below us: George Laven and Fred Dean. My courses lasted about ten weeks but they were there for the one-month short course. Fred was my boss later when I served my time on the Joint Staff in the Pentagon, and retired with three stars. George, who liked fast cars as much as flying, retired as a colonel and went to work for McDonnell Douglas in Israel. We kept in touch, and I learned that between DeLoreans, George was getting some good flying with the Israeli Air Force in F-4 Phantoms and, I believe, F-15 Eagles.

Las Vegas was only six miles away, and affordable. We found we could hire a babysitter, go to town, and see a show, all for about $10. There were other diversions as well. In October Edna and I drove about 150 miles to a spot near Sunnyside for some deer hunting. We arrived about 2100 and were sleeping outside when Edna woke me. "There's something jumping over my face," she said. I told her to relax, "It's just kangaroo mice." That was a mistake, because the next thing I knew she was in my sleeping bag with me. I never did get to sleep, with her fidgeting over the "wild animals."

In late November 1954 I arrived at Field K-3, Pohang, South Korea. There I was surprised to learn I was to take command of Marine Photo Squadron 1 (VMJ-1), a photoreconnaissance unit. I told Maj. Gen. Marion Dawson and his chief of staff, Col. Ed Montgomery, that I hardly knew how to use a Brownie, let alone an F2H-2P. They said, "That's not the reason we're sending you there. We're sending you to straighten out that squadron's problems."

There were certainly problems to solve. After the war was over the previous CO had become more interested in social contacts in Seoul and had allowed things to slide. VMJ-1 had its own little club, although similar clubs were run by the wing and the group, both of which ran according to schedule. But the squadron's club stayed open as long as there were customers. I told the officers they had to close at the same time as the other establishments, but the pilots sneaked it open again. Consequently, I closed down VMJ-1's club and locked the door.

But there were other problems that posed even greater leadership challenges. The pilots were running their missions on a makeshift schedule, almost any way they pleased. They had been flying recon missions up the northeast coast of Korea, staying three miles offshore, looking more for shipping than for anything on land. They would then swing seaward and return over open water.

I decided that wasn't how I wanted to do things. I told the pilots to approach from out at sea and come back along the coast in case they encountered any opposition. But I learned that they flew the next mission as usual, so I called in the flight leader and grounded him. He had been selected for major and I wanted to hold up his warrant, but I was told that wasn't possible unless I put him up for a general court martial. That didn't make much sense to me, so the situation stabilized. And more importantly, the pilots finally learned they had a CO with some experience (this was my third command) and that the squadron would be run my way.

About the only form of recreation at K-3 was hunting pheasant and duck. There were a few deer but I never saw any of them. However, we had wild game feeds at the air group level. Commanding MAG-33 was Col. Soupy Campbell, a real fine person. We hunted together quite a bit, and that helped break the monotony. The war had been over for a year and a half, and things were dreadfully quiet. Another way of easing the boredom was to engage the group's FJ-2 Furys in mock dogfights. It wasn't what photo birds were supposed to do, but it did help keep things interesting.

However, by the end of January 1955 we finished moving the squadron to Atsugi, Japan. I had been at Atsugi for just a few days when I had an accident. I was landing in a slip and held the slip just a little too long. When I flattened out I sank too much and hit the lip of the runway with the starboard wheel, which folded up. I screeched down the runway on the port wheel, nosewheel, and right tip tank.

There wasn't much damage, but the Banshee had to go to overhaul and the whole episode was rather embarrassing. Jets running up at the end of the runway had blown away most of the dirt and gravel from that end, exposing a six-inch lip of concrete. A few days later I got a message from Major General Schilt, Chief of Air, Fleet Marine Force Pacific: "Marion, what you need is a runway stretcher." That's all it said.

On 4 May the routine suddenly changed. I was called to a conference at Yokosuka with Major General Dawson, the Marine Corps wing commander, and Adms. Dan Gallery, Apollo Soucek, and Edgar Cruise. They said VMJ-1 might be assigned a special task on Taiwan, and at noon two days later I was told to fly there and report to Seventh Fleet. I took off at 1630 with a wingman, First Lieutenant Peterson. We rendezvoused at Kadena, Okinawa, and took off next morning for Taipei. En route we ran into heavy weather and, in an effort to stay on top, we ended up at 50,000 feet. Up there we didn't have much to play with. At that altitude in an F2H the difference between buffet and stall is only about twenty knots.

I got a call from Lieutenant Peterson, flying my right wing, saying he was having trouble breathing. I suspected he was hyperventilating. Our canopies were frosted over and we could only see out by scratching a spot with our fingernails, so I couldn't really gauge the situation. But I told Peterson to hold his breath for fifteen to twenty seconds and breathe slowly before reestablishing his normal breathing pattern. That seemed to do it, and shortly we began letting down to the airport at Taipei.

Immediately I took a bus to Keelung to report aboard the USS *Eldorado* (AGC-11), flagship of Vice Adm. Mel Pride. He had been my old boss at NATC but now commanded the Seventh Fleet. During the conference, he and his staff explained plans for clandestine photo flights over Fukien Province, China. There were reports of Red Chinese buildups for a possible invasion of Taiwan and we needed hard evidence one way or the other. Ever since the Korean War had started in 1950 I had wanted another combat assignment, but this was not what I had in mind. We were armed only with cameras—no guns.

Next day I flew down to Tainan to look at the facilities, then returned for another conference with Vice Admiral Pride. Then on the ninth we flew back to Atsugi to meet with General Ennis and his Marine Corps air wing staff. We also had a squadron meeting to lay plans for our move. For additional information and airborne support I flew to Naha, Okinawa, on 10 May to confer with Comdrs. Bob Lund (an old flight school classmate) of Patrol Squadron 29 and Adrian Perry of Aerial Warning Squadron 1 (VW-1). I got into Tainan at 1700 just as six R4Q "Flying Boxcars" arrived with VMJ-1's personnel and equipment.

Next day we sent out our first flight, two Photo Banshees to Fukien Province. Normally I would have led the mission, but I had to return to Taipei to meet Admirals Pride, Felix Stump, and John Dale Price of CinCPac. I was back at 1600 and learned that Major Turcott's flight had met no opposition.

Our Photo Banshees had no guns, so VW-1's Super Constellations were supposed to fly between Taiwan and Fukien Province, remaining over international waters to warn us of any enemy fighters trying to intercept us. There was also radar in the Pescadores, 110 miles from the mainland coast, run by the Nationalist Chinese with English-speaking operators on the scopes. But they weren't completely trained and were hard to understand. Additionally, the Super Constellations' radar wasn't sophisticated enough to break out every target from the ground clutter. Therefore, they just relayed our F2H-2P radio transmissions back to Tainan.

On the twelfth we sent out four flights and I took one myself. We had been assigned an altitude of 40,000 feet by Seventh Fleet, owing to the scale the staff wanted. But that put us into the contrail level, which deprived us of any hope of surprising the Chinese. My wingman and I were probably inland only thirty to forty miles, stretching our necks in all directions, when all of a sudden I saw two MiGs break into the contrails back at seven o'clock some 5,000 feet below. I promptly put my nose down, turned into them, and met them head-on.

As soon as I passed the MiGs I flipped over into a split-S, popped my speed brakes and went straight down, right to the deck. The old F2H was really bucking. I leveled out and went scooting across the terrain, barely over the treetops and mountain ridges. As soon as I got squared away I looked back, expecting I had lost my wingman, but he was tucked in so tight I thought he would rub some paint off my airplane. Back on the ground I mentioned how close he'd been and he replied, "Colonel, I wasn't about to lose you over Red China!"

Although we got away clean, I immediately protested our situation to Seventh Fleet, and requested fighter escort. Admiral Pride said, "I can furnish you some airplanes but I can't spare any pilots." So he sent me four F2H-2s, regular Banshee fighters. I was the only fighter pilot in the squadron and we had no ordnancemen, so I had to teach my guys the basics of air-to-air gunnery and organize a little ordnance section to arm and clean our 20-mm guns.

Next I returned to Keelung and talked to the chief of staff, a blackshoe who had ordered us to remove all identification from our persons and our Banshees. It was an absurd notion from the practical aspect that anybody could identify us by our aircraft, and certainly none of us resembled the locals across the strait. But more to the point, the order would put us in the status of spies if we were shot down and captured. So I talked to Admiral Pride, who was unaware of the order, and he promptly canceled it. While there I went further and said, "Admiral, all I want to know is the scale you need from us. Let me pick the altitude. At 40,000 feet we're in the contrails, and in certain weather conditions they can see us for 100 miles. The other thing is, at 40,000 feet our operating envelope is too restricted. We only have about thirty knots to play with; if you slow down you risk stalling and if you go faster you get into buffet." From then on we went in at 30,000 feet.

Additionally, Seventh Fleet put its best fighter director on the Makung Island radar. He was Lt. Comdr. Bill Hardy of the USS *Corson* (AVP-37), a small seaplane tender. Not only was Bill a top FDO, he was an ace besides. He solved the language problem that existed with the Chinese controllers. I told him, "Don't try vectoring me, I'll do that myself. Just tell me where the enemy are and I'll avoid them." After that we were never intercepted. We were out of the contrails, at a good performance altitude for our airplane, and we had a good guy on the radar.

I didn't learn Bill's side of the story until we met at the American Fighter Aces Association meeting in 1983. He had a powerful new radar on Makung Island that could "skin paint" a fighter-sized aircraft at 260 nautical miles, without transponder assistance. Our Banshees had had their IFF transponders removed for security reasons, should one of us get bagged, but Bill had no trouble separating us from the MiGs. They usually showed on his screen at about 20,000 feet, and with a UHF transmitter he kept us well informed of the bandits' positions relative to us. Bill kept his comments to an absolute minimum; no call signs and only simple code words to indicate where the MiGs were.

From then on I always flew one of the fighters, but we didn't need them. Our specific mission was to look for evidence of a buildup of Red Chinese forces to reinforce the islands of Quemoy and Matsu in preparation for an amphibious assault on Taiwan. But that time of year the weather was poor and we couldn't get many pictures. There never was evidence of a serious buildup, so the program was canceled on 12 June after about seventy-seven sorties.

Years later I learned some of the details from Bill Hardy. He tells of one particular mission he monitored with his big Makung Island radar—a mission that turned into a three-nation airshow: "Marion was approaching my site inbound, returning to Tainan; the Nationalists had four F-86s just north, outbound; and the MiGs were inbound, way behind Marion at a higher altitude. The ChiNat controllers, apparently thinking the F2Hs were hostile, frantically tried to intercept Marion. But since the Nationalists vectored their 86s by Morse Code, it took a long time to order a left 180-degree turn, and the Sabres ended up beneath the inbound MiGs, between Makung and the mainland. I heard Marion's flight pass overhead so I ran outside to watch him heading into Tainan, the MiGs retiring in disarray and the 86s stolidly heading back from whence they came."

Over the next thirty days I took off eleven more times to overfly China, but we had problems with the weather. On six consecutive missions I either aborted at the coast or found the target area obscured as far as 120 miles inland. However, during early June we got results. On the sixth I spent fifteen minutes over the target, flying a fighter Banshee as escort to the "photo joe," and we returned with good pictures. On my last mission, 12 June, we flew around for a full twenty minutes in poor visibility before heading home. Next day we were ordered to halt all operations over mainland China—the crisis had passed.

Ironically, the closest I came to getting shot down was while flying an L-5 over Taiwan. A U.S. Air Force officer, Lt. Col. Clayton Isaacson, used the L-5 for liaison work with the Nationalists, and I borrowed it a couple of times for flying around the countryside. I was never too proud to fly anything, including a Stinson puddle jumper, but on one occasion I was motoring toward Half Moon Bay when flak bursts exploded just ahead of me, at my altitude. I found I had entered a restricted area where Chiang Kai-shek lived!

In June I was pulled out of VMJ-1 and became operations officer of MAG-11, where I flew everything in the group, including FJs and F3D Skyknights. But I returned to Quantico in early January 1956 and by month's end I was at El Toro as the wing safety officer. There I attended a ten-week aviation safety course at USC that caused me to miss one of the worst accidents the air station ever had. An R5D transport crashed near Oakland with thirty-eight dead, and Joe Renner covered the scene for me.

I was never sure why I was stuck with the job of safety officer. I suspected that Headquarters figured if I couldn't help anybody else, at least maybe the course would help me since I had recorded several accidents in my career.

As fate would have it, the next accident involved a close friend. On 16 February a B-52 from Castle Air Force Base crashed near Merced. It was the first Stratofortress lost, but that wasn't what grabbed my attention when I read about it in the newspaper. The instructor, Col. Patrick D. Fleming, was sitting between the pilot and copilot and therefore the only one aboard without an ejection seat. During the flight, one of the high-speed generators let go and punctured a fuel tank. About four inches of fuel was sloshing around in the fuselage, and by the time Pat could make a manual bailout his parachute was afire. He was found facedown in a plowed field, killed on impact. Pat had had a fine future in the Air Force; at the time of the accident he was a colonel and already one step ahead of me, whereas at Pax River he'd been one step behind.

There's a postscript to the story. Late in his life Groucho Marx's companion was an uncommonly good-looking woman named Erin Fleming. Butch Davenport reminded me that she was none other than the beautiful little girl who had lived next door to us at Patuxent River, Pat's daughter.

In April I was tested in a centrifuge for G tolerance and went to 7.5 Gs before passing out. That was without a G-suit. If I needed more tolerance I had to wear a suit, but sustaining heavy G never was a problem for me. Physiologically that is unusual for tall people. The best G tolerance is in relatively short persons because blood doesn't have as far to go from heart to extremities. An Air Force flight surgeon once said that if he were looking for the ideal fighter pilot he would choose a short, overweight major who smoked. Majors have been flying long enough to build a fair amount of G resistance, and overweight smokers have high resting pulse rates, which elevate blackout levels. I prefer my method—inherited G tolerance!

I was determined to finish number one in my safety class, but it didn't come easily. There were some bright boys there, including a couple of Navy lieutenants. I studied even harder than I had in college, and on graduation day in May I stood first. The bright spot in the course was a civilian instructor named Jim Nielson. We became good friends and I took him with me on some local flights.

My tour as wing safety officer wasn't too long, fortunately, and was relatively uneventful. There was time for other activities, including camping trips with the family. That August we went to the Marine Corps Cold-Weather Center near Bridgeport, California. Actually we were between Wolf Creek and Silver Creek, staying in a deserted sheepherder's cabin while I did a lot of fishing. Next morning the Basque sheepherder showed up, along with 700 sheep. I asked if he wanted his cabin back and he said, "No. Too many rats."

Edna overheard the remark and nothing would do but to move to a recreation camp nearby, run by Marine Corps Special Services. It was well equipped with amenities and horses. I think that was the first time Edna had been on a horse—quite a change for a Brooklyn fashion model, but she seemed to enjoy it.

Deer hunting was possible at Camp Pendleton, between El Toro and San Diego. That fall Don Foss (Joe's cousin) and I shot a couple of deer on the eastern portion of the acreage, too far out and too rough for transport by jeep. So I hustled back to the base, grabbed a Piasecki HUP-1 and landed near Don. We loaded our game into the helo and flew it back home. Just a couple of years later that would have gotten us into trouble as a real boondoggle, but it wasn't entirely unprecedented at the time. Fortunately, nobody was around when I landed, and I got the carcass into the car trunk before anyone showed up.

A similar event involved Lt. Col. Hap Langstaff, a neighbor at Tustin. Hap joined us in Mitchell, Oregon, for deer season one fall, commuting from the East Coast in a T-28. Hap bagged a buck, wrapped the carcass in a vinyl sheet, and stuffed it in the Trojan's baggage compartment. During a refueling stop at NAS Olathe, Kansas, the line chief noticed a fluid dripping from the baggage door and fingered it. He asked Hap, "Colonel, what's this? It isn't hydraulic fluid." Hap couldn't resist. He shrugged and replied, "Oh, that's blood. I have a corpse in there." Then he manned up and taxied out, leaving behind a sailor who undoubtedly believed everything he'd ever heard about marines.

I finally made full colonel in October 1956 and became commanding officer of MAG-33. It was almost as good a job as squadron commander, and had the advantage that I could fly every aircraft in the group, including F4D Skyrays and F9F Cougars.

Bill Kellum was exec of MAG-33 but had been passed over for colonel. It was a potentially awkward situation because I had been junior to him one day and now was not only senior but his boss as well. But Bill is a very fine person, and to this day he's never shown any resentment. He was a real fine exec for the year I was there, and we're still in touch. Not many people will take that situation as graciously as he did.

Corky Meyer of Grumman gave me a neon red flight suit that I wore once in awhile, just for fun. And I mean it was garish—before long, word got around that the group CO was the one in the red suit. Well, one day I was taxiing out in an F4D just to log some time when I saw a mess of contrails down toward San Diego. Somebody was having a hell of a good dogfight. I lit the afterburner and piled right into the melee. The "Ford" climbed like a rocket and was fairly maneuverable, so I was having a terrific time. But the F4D was short on fuel, so I broke off early and returned to El Toro.

Word soon arrived that the CO of MAG-15 had learned some of his pilots were involved in the unauthorized hassle and put them on report. Neither the CO nor his exec could fly an airplane very well, and I didn't think much of either one of them. Although the MAG-15 skipper was senior to me, I called him and said, "Look, those guys were just defending themselves, and I was right in the middle of things. If you put those pilots on report, you'll have to cite me, too." No action was taken by him or anyone else. Even if I had wanted to deny my presence, I would have been "made" by my red flight suit. That was the last time I wore it.

During March 1957 the Navy held its last-ever aerial gunnery championship. Edna, Ruth Langstaff and all the kids accompanied Hap Langstaff and myself to El Centro where the Fleet Air Gunnery Unit (FAGU) sponsored the annual event. Hap's VMF-314 won the air-to-air phase, barely edging out VF-51 under Comdr. Alex Vraciu. Alex won the individual air-to-air trophy, which surprised nobody. He was tied with Pat Fleming as the Navy's fourth-ranking fighter ace with nineteen kills, and we'd both been at Pax River in 1945.

In retrospect, the Navy made a serious error in closing FAGU because gunnery was denigrated as a result. By the time things heated up over North Vietnam eight years later, only the Crusader fighter squadrons retained gun armament for aerial combat. They shot down six MiGs for each loss, while the missile-armed F-4 Phantoms did only half as well.

On 18 May I took off in a Lockheed TV-2 for Quantico but had to land at El Centro when a tip tank refused to transfer fuel. I continued to Dallas and then flew on to Quantico, but I wasn't getting any more fuel from the tip tanks. Over Gordonsville I shut down the engine and coasted the rest of the way, intending to make an air start and normal landing, but two things were against me. I couldn't restart and the field was below minimum for visual approaches. There were breaks in the clouds so I elected to make a deadstick landing. Unfortunately, it was raining with low clouds and water on the runway with impaired visibility. In some ways it was worse than the forced landing at North Island a few years before.

Quantico's runway was only 4,200 feet long, with both ends against the river. Even so, I could have gotten away with it if there hadn't been water on the runway. I didn't have any braking so I overran the far end and broke off the nosewheel. So much for being top student in my safety class!

Back at El Toro in June, I flew a T-28 to George AFB to see George Laven. It was about the third time I had visited him, and he made arrangements for me to check out in the F-100 Super Sabre. Chuck Yeager was one of the squadron commanders. He personally oversaw my cockpit check and then chased me on the flight. I don't recall any particular impressions of the F-100, but George's hospitality and Chuck's company were both memorable. It was just one of the many appealing aspects of a career in military aviation.

I passed 9,000 hours in July 1957, enjoying the varied flying, but had a rather sad forty-second birthday on 1 November. On that date I was relieved as CO by Chick Quilter and became wing operations officer. In my first month as G-3 I flew only thirty-eight hours, although tactical aviators today would give their right arm for even that much flight time.

That summer Edna, the kids, and I prepared for our next move: Maxwell AFB in Montgomery, Alabama, and the Air War College.

CHAPTER TEN

War College and Washington

The Air War College class of 1958–59 boasted five marines, including one ground officer, Glenn Long, and four aviators: Art Adams, Paul Ashley, John Howard, and myself. And our family soon expanded. Edna's mother stayed with us in extra quarters I had arranged, and she remained with us until her death four years later. Butch Davenport flew cross-country in his Navion to deliver our seven-year-old son Bruce from Oregon, so the whole family was together again. And almost immediately after the class arrived in August, Art Adams's wife, Katie, gave birth to a baby girl.

Despite the academic nature of the Air War College, I found it an enjoyable tour. One of the first things I did was to arrange a check ride as an instructor pilot in one of Maxwell's T-33s. I already had quite a bit of time in type from the early Navy P-80s and TV-2s, so I was soon up to speed. Additionally, I got an IP rating in the C-45 to keep current in reciprocating aircraft types. The dual rating worked well at the Air War College, as no other high-level school would have provided me with as much flying opportunity. I also gave rechecks to many of the students who had flown the T-33 earlier in their careers.

We had a good bunch of people in our class. I gravitated to the fighter pilots, of course; especially aces like Reade Tilley of Eagle Squadron fame and Billy Hovde, who had flown Mustangs in Europe and F-86s in Korea. We had an excellent series of speakers from government and the military: General White, the Air Force Chief of Staff; Gen. Earl Partridge of NORAD; and Alan Dulles and Henry Kissinger. Joe Foss even dropped in unannounced a time or two.

In 1955 Joe had become governor of South Dakota, and in his first executive decision he fired his pilot and sold the executive aircraft. Joe said that if there was any flying to be done, he would do it—in a South Dakota Air National Guard fighter! Most people assume his brigadier general rank came in the Marine Corps, but it was actually with the Air Guard, where he logged 1,600 hours in P-51s.

About 0100 one night Edna and I were awakened by pounding on our door. We ignored it for awhile, as there was always somebody wanting to party into the wee hours. At length the noise faded. Minutes later there was a rapping at our bedroom window. It was Joe Foss and Col. John C. Meyer, one of the staff instructors. They were in rare form. It was no surprise in Joe's case, since he loved to talk to old friends, but Johnny Meyer was usually a rather dour sort. Although he had been a top ace in Europe and got a couple of MiGs in Korea, he had never developed a reputation as a partygoer. Maybe Joe was an undue influence. Anyway, they said that if we didn't join their party they'd bring the party to us. We decided to accept their invitation.

That War College class was always ready for a get-together. Sometimes you'd come home in the afternoon and you'd find your neighbors in your living room, drinking your booze and wondering where you'd been. I don't know how I found time to write my thesis on limited wars. Eventually Edna and I took night classes three times a week—investments and ballroom dancing, among others—just to escape the constant round of parties.

But it wasn't all socializing. There was nearly a tragedy when the class attended a firepower demonstration at Eglin AFB, Florida, in early 1959. Our family and the Adamses were enjoying the beach when little Melissa Adams wandered too far into the surf and was swept off her feet, badly in danger of drowning. Edna sprinted into the water ahead of everyone else and reached Melissa, but almost immediately she got into trouble herself when the little girl became panic-stricken. Art and I had been fishing farther up the beach, and by the time we rushed out to the surf it was almost too late. Thanks to the young sons of an Air Force couple named Dixon, Edna and Melissa were pulled in to shore. But we nearly lost them both.

My next duty wasn't nearly as satisfying as Maxwell. I had requested to relieve John Condon as the Marine Corps representative at Commander-in-Chief Europe, but that spot was already filled. Instead, I went to the Joint Chiefs of Staff in Washington. That June we located a house in Arlington, Virginia, and I reluctantly entered the world of Washington politics.

Our next-door neighbors were Rep. Charlie Brown of Missouri and his family. They were delightful people and we saw quite a bit of them. And at first the JCS job wasn't too hard to take. I reported to J-3, the European Division of Plans, as one of the naval representatives. It didn't take me long to discover the amount of work wasn't very demanding, and I was able to get in a reasonable amount of flying. However, that was to change with the new year.

In November 1960 John F. Kennedy narrowly won the presidency from Richard Nixon, and America began its love affair with the new first family and the glitter of "Camelot." Despite all the grief that followed, the enthusiasm was understandable. Following Truman and Eisenhower, the forty-two-year-old Kennedy and his youthful entourage were a refreshing change from traditional presidential families. But the pack of advisers and bureaucrats that swept into office with his victory soon brought lasting problems to military professionals. Those individuals became burdensome baggage we had to pack around long after Camelot had faded; not the least of these was Secretary of Defense Robert McNamara.

Working with me in the J-3 office were Cols. Bill DeLacy of the Air Force and George Barker of the Army. George had been on NATO staff in Norway and had married a local girl while there. He had a lot of staff experience, and knew more about handling Department of Defense staff work than either Bill or I. Often we were asked to review a situation and comment on it, so usually George would dictate our reply to a stenographer and we'd be done for the day. It was even possible for Bill and me to go hunting or flying for a day while George manned the office.

Then the McNamara regime took over at the Pentagon. He began to shove a lot of work onto the Joint Staff, mostly in the form of studies. We'd send a study over to SecDef and it would come back marked "Disapprove, restudy." In some cases we went through the evolution a half-dozen times. It took us a while, but gradually we realized that McNamara and his whiz kids—none of whom had any military experience—already had their minds made up. It didn't matter what the subject was because they weren't interested in the professionals' opinion or the benefit of our experience. SecDef could change a board's findings at the stroke of a pen, saving us the task of arriving at a politically correct solution. Although they never actually said so, the whiz kids merely wanted us to rubber-stamp their own preconceived notions. Thus, we lent credibility to some pretty off-the-wall concepts. It was a galling situation, but we couldn't argue or escape.

Finally, we worked it out. By going through the back door we'd learn what McNamara really wanted to justify. Then we would send a study that met his criteria, thereby saving ourselves much effort that could be devoted to other tasks.

Everybody said McNamara was a certified genius at management, and maybe he was when at Ford. But the managerial methods of big business simply don't apply to military operations, and apparently he and his crew were unable to make the distinction. One example will suffice. The TFX concept was a multipurpose tactical fighter, intended to be all things to all aviators. Anybody who knew anything about military aviation realized the idea was ridiculous; any airplane is necessarily a compromise with emphasis in one direction or another. But McNamara's band insisted that what became the General Dynamics F-111 could equally perform fighter and attack missions for both the Air Force and the Navy. We called it "The Flying Edsel."

With a variable-geometry wing and powerful engines, the TFX was to have supersonic performance *and* long range. We were told it would be able to fly several hundred miles to its target, loiter near the objective, and return. When some of us tried to explain that loitering in the target area would make the airplane a fat, tempting target, McNamara's answer typified ignorance: "Loiter in afterburner." Where all that extra fuel was to come from no one could say.

Eventually, Vice Adm. Tom Connolly, DCNO(Air), with help from sympathetic members of Congress, was able to kill the proposed Navy F-111B. The B model was simply too large and too heavy to operate from aircraft carriers. In testimony before Congress Connolly said, "There isn't enough thrust in Christendom to make that airplane into a fighter." However, much of the technology developed for the F-111B turned up later in Grumman's F-14, and the Navy gratefully named the new fighter "Tomcat" in recognition of Tom Connolly.

In the end, McNamara proved to the country and to the world that he was completely unsuited for the position Kennedy had chosen for him, and the man inflicted lasting harm upon the U.S. military. Perhaps the most succinct assessment came in *The Best and Brightest* when David Halberstam said of McNamara, "He was, there is no kinder or gentler word for it, a fool."

In addition to growing concern about the military's civilian leadership, I had a personal concern about an old friend. In January 1960 I went to Bethesda Naval Hospital to see John L. Smith in the psychiatric ward. It was a rather pathetic situation.

John L. had been perhaps the strongest squadron commander and finest combat leader I had ever known. He had come to Washington as a colonel in the number two spot in the Division of Aviation. Based on his previous record, everyone figured he would easily make brigadier general. Unfortunately, while at Marine Corps Headquarters he alienated some of the ground officers. He was so pro-aviation that he didn't take into consideration that he had to get along with the ground. Since the selection board was six ground officers and three aviators, that was the end of his chances for a star. He was a proud type, and getting passed over hit him pretty hard. Consequently, his career in the Marine Corps was effectively over.

Other episodes from that Washington tour were much more pleasant. Marines always enjoy showing the Navy how to do things, and in the summer of 1960 I had an interesting episode with Jerry Miller, who later retired as a vice admiral. He lived near us, and one day he mentioned that he was going to hire a professional to remove three trees on his property. He was afraid they might fall on his house. I jumped at the opportunity and offered to take them down in exchange for half the wood. He said, "It's a deal, but I don't know how you're going to do it."

I borrowed a pair of pole-climbing spikes from Quantico and climbed each tree, topping them as a preliminary move. Then I tied ropes to the trees near the top and used the ropes to guide the trees away from the house. It was an incident I never forgot, and neither did Jerry. He always mentions it when we meet, and asks if I'm still climbing trees—which I do.

In July 1961 I checked out of JCS and moved to Marine Corps Headquarters as number three in the Division of Aviation. Shortly thereafter I was pleased to learn that Keith McCutcheon would take over as Director of Aviation, even though he was still a colonel. Keith was smart and articulate, the ideal officer for the job, and I was pleased to work as his deputy.

The new job meant more social obligations—something Edna always enjoyed far more than I did—and one episode proved memorable. In September we attended a reception at the Commandant's house, sponsored by Gen. David Shoup. It was formal, of course, with all the officers dressed in whites. Shoup walked up to me and said, "I have a present for you." We went up the steps and he addressed the entire reception, about two hundred people, gathered on the lawn. After a rather flowery speech (he wrote poetry in his spare time), he presented me with a pair of shower shoes he had taken from beneath my cot on Guadalcanal nineteen years before when I had been shot down and was missing. He said he considered them a good-luck charm, but since he didn't think he would need them anymore he was returning them to me.

Shoup was not one to drink heavily, but in this case I think maybe he'd had one too many. I suspected he'd only meant to give me one shoe, keeping the other as a souvenir, so eventually I returned them to him. Evidently they meant something to him, and I wanted him to keep them.

Dave Shoup was a complex individual—smart and tough, with a streak of the philosopher in him. He was also known as a brutal card player who took no prisoners. Rumor had it that his 2,000-acre farm in West Virginia had been financed largely by his poker winnings.

Later it occurred to me that I may have unwittingly made a high-powered friend. In January 1962 Keith announced he was going to command the 1st Marine Expeditionary Brigade, having been selected for brigadier general. A few days later I was called to the Commandant's office, where Shoup indicated I would be the new Director of Aviation.

My appointment became effective in February, just in time to attend a SecNav dinner honoring another Marine Corps aviator, John Glenn, who had just become the first American to orbit the earth. Some officers disapproved of John's political ambitions and his close ties to the Kennedys, but no doubt much of it was envy. Of course, later he was elected to the U.S. Senate from Ohio and ran for the Democratic presidential nomination.

About this time General Shoup decided the Division of Aviation title would be abolished and the position would be called Deputy Chief of Staff for Air. This put aviation one echelon above general staff, meaning that as a colonel I could sign off not merely "by direction" as before, but as DCS Air. It put me in a potentially embarrassing position because on the staff chart I was one echelon above Gen. Leonard Chapman (G-4), who later became Commandant, and General Cushman, who was G-3 and who followed Chapman as Commandant. I went down and talked to them about it and they assured me not to worry; it wasn't my doing. I think Keith McCutcheon was the one who persuaded Shoup to take this route.

Although I was only a colonel, Edna and I were invited to several formal functions that normally were attended only by flag officers. We also did some entertaining in return, hosting some fifty guests on a few occasions. I was able to do a fair amount of flying in my five months as DCS Air, mainly in T-28s, and that helped keep up my morale through the bureaucratic rough spots.

One of the flag officer receptions Edna and I attended featured John Glenn escorted by Lyndon Johnson, then vice president and nominal head of the space program. Lyndon was visibly displeased with his role, which was more that of a valet to Glenn than vice president to Kennedy. As Edna and I walked out, Johnson was standing alone in the hallway, and he was mad. Edna made an innocuous comment to him, and he just glared at her without saying a word. I knew what was coming next, because my wife has a hair trigger at times. Her mouth opened and I grabbed her by the elbow, knowing that she was about to let Lyndon have it with both barrels—vice president or not. I dragged her away, and a few minutes later she admitted, "Marion, it's a good thing you did that because you'd probably be a PFC before we got out of there." I've learned to be pretty fast on my feet after living with her for nearly fifty years. But her concern was not misplaced, since Johnson proved to be a petty, vindictive politician who never forgot a grudge.

On 5 July 1962 Brig. Gen. Norm Anderson relieved me as DCS Air and I moved down to the number two aviation spot. A few days later General Shoup gave a reception attended by President Kennedy, affording Edna and me a chance to meet him. It was a brief meeting, hindered by the formality of the occasion, and I retained no particular impressions of the event.

I had a better time at a more informal get-together a few days later. Our new neighbors, Maj. Gen. Al Boyd and his wife, had bought Charlie Brown's home when he and his family moved back to Missouri. General Boyd was retired by then, working for Westinghouse. The Boyds were very private persons and tended to be all business, but we lived next to them for about a year and I had a great deal of respect for Al Boyd. If you put all of Chuck Yeager's experience together with mine, it just about equaled Boyd's. When he retired from the Air Force he had 23,000 hours in 398 aircraft types and models. As far as I'm concerned, he was the greatest and most experienced test pilot the world has ever seen. One of the things I most admired about General Boyd was that he kept flying operational and experimental aircraft despite his rank—something a lot of general officers didn't do even when they had the opportunity.

By late 1962 McDonnell's new F-4B and F-4C Phantom IIs were in service with the Navy and Air Force, respectively. I attended a joint-service conference at Wright-Patterson that October and proceeded to St. Louis to look over the RF-4C. Mr. McDonnell was there to greet me with his old standby question: "You don't happen to be looking for a job, do you?" Although I wasn't, it was nice to be wanted.

As more Phantoms became available the Marine Corps got its initial batch, and in January I first flew the F-4B at Cherry Point. Maximum velocity proved to be 1,150 knots, or Mach 2.05, and I was suitably impressed. Eventually I logged a lot of time in the F-4 and came to regard it as the jet equivalent of the F4U. In their times, the Phantom and the Corsair were the most versatile aircraft the Marine Corps possessed, with exceptional performance and service life. I had been a single-seat pilot all my life, so having the radar intercept officer in the F-4's back seat required some adjustment, but the Phantom was head and shoulders above its contemporaries. I always felt that whatever faults were blamed on it had more to do with the way it was employed than with any intrinsic deficiencies. Marine Corps fighter-attack squadrons flew the F-4S until 1989, and that's a long run by any measure.

That same month I learned I would be going to Hawaii as chief of staff for the 1st Marine Expeditionary Brigade. Carl Youngdale was going to relieve Keith McCutcheon as CO, and I had asked Carl if he had picked his chief of staff. He asked if I was interested and said he'd love to have me. That settled it; we would leave for Hawaii in the summer.

On 16 July 1963 we checked in at MCAS Kaneohe, where we were met by the Quilters. We had been there for less than a week when Chick called to tell me that we had been selected for brigadier general. He was at the top of the list, having been passed over the year before, and Art Adams and I were at the bottom of the list. Things were really looking up. Edna and the kids loved Hawaii, and I was pleased to be back with the operating forces. As for making brigadier general, that wasn't hard to take, but I always felt that Edna did as much as I did to get my first star. She always was extremely popular, and made lasting impressions wherever she went. But I would have to work to earn that next star.

Just before New Year's Eve 1964 the Youngdales went to Maui and I got a call from Commandant Gen. Wallace M. Greene. He said that Carl was going to Saigon on the MAC-V staff, and that I was to relieve him in a few weeks. When Carl returned from Maui he was surprised to learn he was going to Saigon. His next question was who would relieve him. When I said I would, he exclaimed, "I should have known you were going to make BG. Congratulations!"

I was glad to get the brigade but sorry to see Carl and Jean leave. I really enjoyed working for him, and they were a great couple. But now there was a lot to do in not much time. The next war was just around the corner.

CHAPTER ELEVEN

Vietnam

On 24 January 1964 I assumed command of the 1st Marine Expeditionary Brigade and pinned on my star. The family soon moved into general officers' quarters at MCAS Kaneohe, Hawaii, which Edna really appreciated because a steward was available to help her with housework and entertaining. My own work began to pile up.

When I took over the brigade, Lt. Gen. Carson Roberts had Fleet Marine Forces Pacific (FMFPac), and I looked forward to working for him. He was a real gentleman. But I'd had the brigade only a few weeks when I learned Victor Krulak was coming; this was the same "Brute" I had hoisted on my old Sikorsky at that Quantico demonstration so many years before. Frankly, none of us was looking forward to the command change because we knew that Krulak would be more demanding, particularly where the ground forces were concerned.

Krulak relieved Roberts as FMFPac at the end of February. I made a week-long inspection trip to Taiwan, but still managed to squeeze in some good flying. On 9 March I made five landings and five catapult shots aboard the USS *Constellation* (CVA-64), flying with VMF-212's F-8E Crusaders. The ship's captain was Fred Bardshar, a World War II ace and former test pilot. While aboard, I was told it was probably the first time the Navy had experienced a general coming aboard ship while flying his own airplane. Later that month I passed 11,000 hours.

General Krulak liked to put on demonstrations, and my brigade was the only Marine Corps unit in Hawaii that could accommodate him. During demonstrations he usually looked over my shoulder, and rightly so, since it was my first experience commanding ground troops. I had an entire regiment with a couple of additional battalions, plus an aircraft group of three squadrons. Other than occasional comments, the Brute never said anything to me about aviation, but I picked up a lot of flak about the infantry!

However, I think I proved that I could handle "grunts." In May I observed an exercise at Pahakaloa, a joint training area with the Army on the big island of Hawaii. While I was discussing our defensive setup with the ground commander, I remarked that a certain area did not seem well defended. I had shot a ram in that area a couple of weeks before and packed it down from 9,000 feet off Moana Kea. My infantry leader said, "General, the terrain is too rough. They'll never come through here." Guess where I was standing at 0200 when the aggressors came through our lines?

That spring and summer was an exceptionally enjoyable period. I took a scuba course at Kaneohe and did some diving, and Bruce and I went to Sea Life Park to swim with the porpoises. It was really a lot of fun. Then in July we were invaded by Otto Preminger with John Wayne, Kirk Douglas, and Patricia Neal for the filming of *In Harm's Way.* Edna was thrilled with this development because we had the Douglases at our quarters for lunch and could watch some of the shooting. We particularly liked Mrs. Douglas. However, I'm sorry to say we didn't develop a lot of enthusiasm for John Wayne, despite his sympathetic portrayal of marines.

I tried to keep current in each of the aircraft in my command, and on 13 July I shot ten landings aboard the USS *Kearsarge* (CVS-33) in an A-4C Skyhawk. A bit later I requalified in the UH-34 helo aboard ship with five day and six night landings. Also, I "bounced" an F-8E Crusader in field carrier landing practice. I intended to get night-qualified, but somehow it never worked out. These evolutions proved worthwhile. On 5 August the brigade was put on notice following Viet Cong attacks in South Vietnam and the Tonkin Gulf incidents.

On 7 March 1965 *In Harm's Way* premiered in Hawaii. The title and the timing were appropriate. Two days later we received orders to deploy to Vietnam.

Edna saw us off the sixteenth when the staff and I boarded a C-54 for Kadena, Okinawa. My chief of staff, Col. Regan Fuller, reported that unloading was ahead of schedule, so on 2 April I went to Danang to check on our helicopter elements. A series of conferences followed with senior Marine Corps officers and General Westmoreland, the Army officer in overall command of U.S. forces in South Vietnam. Then it was back to Okinawa.

But not for long. About midnight on the fifth I was awakened and called to the command post. Before noon I was wheels-up for Subic Bay in the Philippines as Commanding General, 3rd Marine Amphibious Force (MAF), with my nine-man staff. We boarded the USS *Mount McKinley* (AGC-7) and sailed at 1700 that same day. Three days later I went ashore at Danang by chopper for lunch with General Westmoreland and several others, then took a helo down to Phu Bai for an inspection before returning to the ship.

On the tenth two companies of 2nd Battalion, 9th Marine Regiment choppered ashore into Danang. I took another helo to Hue and returned to the *Mount McKinley* by Army boat via the Song Hue River on the thirteenth. However, I had satisfied myself that Phu Bai near Hue would be the base for our chopper squadron. By the fourteenth nearly all of 3rd Battalion, 4th Marine Regiment was ashore so I went upriver by gig. More meetings ensued, then I hopped a C-130 back to Futema, Okinawa, to oversee the rest of the brigade's move.

I took out some additional insurance while in Futema. I already had a government-issue .45 automatic, but I went to the PX and bought a .44 magnum revolver as well. I figured that personal weapons are like antifreeze; you can only have too little—never too much.

On the twenty-ninth I was back at Cubi Point aboard another command ship, the *Estes* (AGC-12), and we sailed for Chu Lai on 2 May with most of the same staff I had had for the Phu Bai landing. D-Day was the seventh.

After my inspection of Chu Lai I concluded that the beach had the softest, finest sand I had ever seen. I told the staff we would have trouble getting the wheeled vehicles ashore and suggested that the matlayers should be offloaded first so they could put a carpet down over the sand. But due to some confusion in combat loading the vehicles, we were unable to send off the matlayers first. The first vehicle ashore was a communications jeep, and it bogged down, even in four-wheel drive. We had to fuss around to unload a bulldozer next to make way for the matlayers. Unfortunately, it was not unusual for some ground officers not to take the comments of an aviator seriously. If there had been an opposed landing the Viet Cong could have swept us off that beach with a broom. As it was, our greeting came from a group of Vietnamese flower girls.

There's a story that may or may not be true, but it was said that Chu Lai never existed as a place-name before the Marine Corps arrived. Reportedly it was the closest the Vietnamese could come to pronouncing "Krulak," and certainly Brute did nothing to quell the rumor.

My initial relations with the South Vietnamese Army (ARVN) were good. I visited General Lao's 2nd Division, and we worked with them fairly closely from then on. However, the relationship was conducted under new terms in late May because my command—3rd MAF—was taken over by Lt. Gen. Lew Walt, who had been jumped over twenty-two other officers to win his third star. I returned to Futema a couple of days later, then proceeded to Iwakuni, Japan. It had been a year since my last leave, so I called Edna to join me in Japan until I became deputy commander of the 3rd Marine Aircraft Wing under Keith McCutcheon in late August.

Edna and Lyanne came out by ship, and about ten days later Lyanne returned to Hawaii with friends. While at Iwakuni Edna said she wanted to see the nightlife, so I agreed—with some trepidation. I knew we would run into Marine Corps officers and enlisted men with Japanese girls, and said so. She replied, "Oh, that's all right, don't worry about it."

Touring with my new chief of staff and aide, Col. Tom Bronleewe and Lt. Joe Maiden, we visited some bars without incident and finally came to Harry's 500 Officers' Club. Tom and Edna got out of the car and went into the club while I discussed some matters with Joe for a moment. We were just getting out when Tom and Edna came back, both giggling. I asked what happened and they said, "We were thrown out. No round-eye women allowed, only Japanese females!" I was about to charge inside but thought better of it. Instead, next day I had Tom inform Harry that he had a choice. He could remove the sign indicating it was an officers' club, or he could permit an officer to enter with any guest. If he did neither I would declare the club off-limits. He removed the sign.

On 30 August I joined the 3rd Marine Aircraft Wing as deputy commander and settled in at Danang with all my gear. I intended to fly as much as my duties allowed, and that proved to be quite a lot. I was up to speed again in the Phantom, and a week after arrival I went down to VMFA-542 and logged my first combat mission in Vietnam, dropping six 250-pound bombs north of Hue.

About this time I got a visit from Dick Mangrum, now a lieutenant general at Marine Corps Headquarters. I always enjoyed seeing him, and we couldn't help comparing notes about how different things were from Guadalcanal. There were other old friends around as well, including Col. Frank Scott, who ran the Air Force side of Danang. He had been a War College classmate. Frank was a real fine guy, and he gave me any help I needed, but he always cautioned me not to let word leak out to Gen. Joe Moore, Westmoreland's deputy, who ran Southeast Asia air operations. "If he hears I'm helping you, I'll be a dead duck," Frank said. That was usually how it was. The Army, Navy, Air Force, and Marine Corps got along well at the tactical level, but at higher echelons they were always protecting their turf, roles, and missions.

One example was the continuing "sortie count war" between the Air Force on one hand and the Navy and Marine Corps on the other. The various services were gripped by the McNamara regime's quantification fetish, in which numbers were everything: body counts, sorties flown, bombs dropped. It wasn't unusual to see four F-100s or A-4s taking off with a single 250-pound bomb below each wing, even though the mission's entire ordnance could be carried by a single plane. It was part of the interservice rivalry that characterized so much of the U.S. effort. On 16 September, for instance, I made a radar-controlled bomb run on instruments west of Chu Lai, dropping four 250-pounders from 20,000 feet. My wingman in his

F-4 had a similar loadout. Two Phantoms delivered 2,000 pounds of ordnance when just one could have dropped twice as much.

Johnson and McNamara didn't seem to mind such shenanigans. All of us at Headquarters called them "Lyndon" and "Bobby," without an ounce of respect or affection for either one. There were lots of other Washington types willing to drop in, be seen, take up our time, and generally get in the way. I began to think it was one hell of a way to fight a war.

One of the few who made a favorable impression was Senator Brewster of Maryland. I took him out in a gunship and he wanted to get right in the middle of things. Of course, I had been cautioned by Lew Walt and everyone on down not to take a visitor into a combat area, but I fudged as much as I could for Brewster and he got to see a little firing from a distance.

In late October I flew gunship escort on a sightseeing tour for Senators Kennedy and Tydings with Representatives Culver and Tooney. Of all the politicians and bigwigs who came out to Vietnam, Teddy Kennedy impressed me the least. He had a bunch of press people following him around and making a big show, but far as I could tell he didn't give a damn about what went on as long as he made the news.

One incident that particularly incensed me occurred when Kennedy visited "B Med," the makeshift medical company that received wounded marines from the front lines. The place was immediately saturated with photographers and members of the senator's party, and in jostling for position they banged into bunks occupied by seriously wounded marines. If I'd had the authority I would have evicted the whole bunch.

But visitors were the least of our worries. The locals gave us plenty to think about. On 28 October the Viet Cong hit Marble Mountain chopper base in the Danang complex with an extremely effective rocket attack. We had to write off seventeen UH-1Es and five H-34Ds, plus two A-4Cs at Chu Lai.

I did a lot of flying in a Huey with four .30-cal. guns—what we called a "dirty slick." During two days in late November I logged five medevac flights through miserable weather following heavy attacks on ARVN positions at Hip Duc, Tam Ky, and Ki Ha. The Viet Cong had attacked Tuc Ca at 0400, and when I landed there on the twenty-second a couple of hundred dead were lying around. I lifted three loads totaling twenty-one casualties in my Huey, and checked with General Lam as well. This was a situation I found myself in more than once. The ARVN said I was the only helo in the area, as low ceilings prevented medevacs and reinforcements from getting in.

On one occasion I landed at an ARVN compound that had beaten off an attack. The South Vietnamese asked if I could take out some casualties. I said yes, but I wasn't paying close attention and they stacked the casualties on top of each other like cordwood. I had a copilot and crew chief plus the four external guns, so we were really loaded. As soon as I tried to lift off I knew I had a problem, but since we were at the top of a steep hill, once I jumped into the air I could lose altitude and gain airspeed for translational lift. We landed at the closest hospital helipad with at least thirteen casualties aboard.

In late November I went back to Iwakuni briefly to interview candidates to replace my aide, Joe Maiden, who had just been promoted to captain. I found Lyle Prouse, one-quarter Comanche from Wichita. A series of bureaucratic bungles had prevented Lyle from gaining admission to Annapolis, and when I met him he was an A-4 pilot in VMA-223. He definitely did not want to be a "seeing-eye lieutenant" for some general, but despite his protests he was one of four candidates nominated by his group commander. I interviewed him and let him go. He thought he was off the hook. Lyle takes up the story:

"About two weeks later I was informed I was to be General Carl's new aide. I couldn't believe it. I had two and a half days' break-in under Joe Maiden, whom I had known at El Toro, and literally ran into Brute Krulak during an inspection. I snapped to, saluted, and stared straight ahead. I didn't dare look down—I'd heard the stories—and I looked right over the top of his head, but I noticed he carried a polished swagger stick. Krulak merely asked, 'Whose aide are you, son?'

" 'I'm General Carl's new aide, sir.'

" 'Well, he's a good man. Take good care of him.' Then he returned my salute and marched off.

"Joe Maiden gave the general top billing but warned, 'He likes to fly an awful lot, but you won't get much stick time.'

"The four months as an aide were almost the only time in seven and a half years on active duty that I was out of a flying billet. But in those four months I logged eighty hours in Hueys, and that was only counting the time General Carl let me fly. Stateside, twenty-five hours a month in a jet was doing real well.

"I became a semiqualified helo pilot strictly by on-the-job training. The first three or four flights I rode in back, but then I found the Huey manual on the general's desk. There was no time to read it but I scanned it and got some buzzwords from it."

During a flight to Monkey Mountain, Lyle asked me about translational lift—the phase when converting from vertical to forward flight. That showed me he had done some background reading on helos, and from then on Lyle was always sitting in the left seat. He recalls:

"Flying down the coast one gray, drizzly day, I looked over and was astonished to see General Carl asleep. I couldn't believe it. I wasn't really a helo pilot, I was just hanging onto the thing. Then a Skyraider broke out of the gloom directly ahead, flying up the coast. I broke hard to maintain clearance and the general came awake. 'I had to clear an AD, General,' I explained.

" 'Oh, OK.' Then he went back to sleep!

"Nothing seemed to rattle him. On one occasion we landed on the edge of a field to eat C-rations, listening to a firefight from beyond a treeline with stray rounds ricocheting nearby. The general continued eating nonchalantly while I resisted the urge to take cover behind something.

"On 12 December we took a round through a rotor blade near Dong Ha. Upon return to Marble Mountain a squadron lieutenant asked me if I had been ordered to fly back with the general. I said no. The lieutenant said that would be the only way he'd climb in a helo with a round through the rotor spar. He asked me—a fixed-wing pilot—if I knew about Dead Man's Curve. (He didn't mean the song by Jan and Dean.) Then he said, 'You'd do well to read up on it.'

"Marion Carl's standard departure was straight up to about fifty feet, then drop the nose and accelerate into forward flight. It was like going up in a real fast elevator. Well, Dead Man's Curve was the plot on the chart where minimum knots were necessary in minimum feet to land safely in the event of an engine failure on liftoff. We always lifted off at Dead Man's Curve.

"There were other things I had to learn about the helicopter war. I remember the first time we rolled in on something, and I couldn't believe how long we were in the run. I was used to being in and out at 450 knots with a four-G pullout, but it seemed we hung up there for about an hour and a half. I thought, 'Hell, if I was sitting down there in that hooch I could hit us with a rifle.'

"I told my friends that General Carl was the last of the old warriors because he didn't spend a lot of time sitting behind a desk. We covered a lot of ground, and usually we were gone all day on business to log three or four hours' flight time. I wish I had a nickel for every time I defended him in some bar after I rotated home. He was controversial because he was decisive, and that usually wasn't how things were done in Vietnam.

"The one problem, however, was that the general hardly ever used a call sign—Moment Six, for instance. The tactical air control folks asked me to sneak a call once in a while 'so we know where the general is.' That was because we hardly ever flew with a wingman."

The new year began well for me. On 1 January 1966 I won $45 playing poker in the Danang Officers' Club. A few days later I got a call from General Kier in Hawaii, who put Edna on the phone. It was Brute Krulak's birthday, and everybody was having a good time at the party.

On the fifth I had another escort for H-34s fifteen miles southwest of Danang. We drew fire twice but weren't hit. We also chased a couple of Viet Cong but didn't shoot them. I was willing to risk a capture in some circumstances because the intelligence value was high.

Lyle was with me both times we captured Viet Cong suspects. On the first occasion, Valentine's Day, we logged nearly six hours supporting Operation Double Eagle near Phu Bai. We were flying down a streambed when a youngster broke cover and the door gunner opened fire. Lyle's recollection is remarkably detailed:

"General Carl called on the intercom, 'Hold your fire, damn it. He's just a kid.' I didn't think it was a kid, but looking out the panel, past the general, I saw the Vietnamese crouching in some tall grass beaten down by the rotorwash. I was excited—my blood pressure must have been 400/300—while the general seemed calm as ever, motioning with one hand for the Vietnamese to come aboard. Finally I thought, 'Damn it, if you'll set this thing down I'll go get him.' Then I realized I had said it out loud! There was nothing to do but round up the Vietnamese, who had stashed his weapon in the water. We delivered him to the Army at Quang Nai."

On another occasion we were supporting an extraction near An Hoa. Rocket-armed helos were beating up a treeline when several people fled across a river during a lot of shooting. I landed on the far side of the river and said to Lyle, "There's a couple of them behind that sandbar." I knew he was pretty eager because sometimes when I was strafing he felt left out and asked to shoot his M-16 out the starboard window. This looked like a good chance to get him some action. I lifted off in a tight 360 so Lyle could see them. Instead of two bodies there was now one—the other obviously had ducked into the brush. Then I asked Lyle, "Do you want to go look for him?" My aide's reaction was nonregulation:

"I wanted to say, 'Are you crazy?' But it came out, 'Yes, sir.' I grabbed my M-16 and jumped out, looking for a way into the thicket.

"I only found an entry after the second or third try. It was early in the day, but the jungle canopy made the area extremely dark. I kept switching from automatic to semiautomatic. I thought, 'If I go to semiauto I'll be too slow if I find this guy.' Then I thought, 'If I go automatic this damn thing will jam on me.' I would switch back and forth. I was scared. I wondered if this guy had a wife and kids, since I had two boys of my own at this time. I wondered where I'd hide if I was wounded and the enemy was all around. I would try to take one with me.

"Then I found his AK-47. The front handguard was shattered and there was blood on it. I did not want to find that guy; I knew that if I did and he moved, I'd shoot him. Otherwise I'd likely catch one through the running lights, and how would anyone explain that to Barbara? If I got killed I was supposed to be in a cockpit, not down here running through the woods.

"When I came back out of there I didn't feel much like a marine because I thought my killer instinct had failed me. I was still scared. But the gunner was at the edge of the woods, yelling at me, 'The general wants you to come back.' I thought that was the best news I'd had in the last twenty minutes, or five minutes. It seemed like an eternity."

When Lyle emerged from the thicket I was on the radio to one of the rocketship pilots. The pilot asked for the AK-47, and I agreed without thinking to ask Lyle. He was really peeved. When he delivered the rifle to the VMO-6 pilot, he said, "Next time you want a souvenir, you can damn well get it yourself." Lyle was going to ask the group commander for it, but I told him, "I'll get you another one just like it." However, as he recalled, "I felt like stamping my feet like a four-year-old, pouting and whining. 'I don't want another one. I want that very same rifle!' " He still reminds me that he doesn't have his trophy AK-47.

In the early part of the year we had lots of visitors, including Robert Mitchum and Charlton Heston. I especially liked Mr. Heston—a World War II airman—and tried to get permission to take him up in an F-4. But I couldn't wrangle it during the one night and day he was there. I also saw quite a bit of Johnny Hyland, who had been assistant director of Flight Test at Patuxent River and was now a rear admiral. Johnny and I got together on the hospital ship *Repose* (AH-16) after I escorted General Westmoreland there on an inspection tour.

In early February Krulak arrived, bringing tapes and fresh fruit from Edna. V. C. Kahn and Mr. Chow, businessmen from Hong Kong, also were in Danang that day. Kahn was quite an operator, widely known in Hong Kong, but for some reason he had a soft spot for marines. I told Edna to look him up when she got there because he would steer her to the right places for the best deals. I saw V. C. again the next month when I flew to Hong Kong as copilot on a C-130 and checked in at the Astor. A cruise ship, the SS *Sagafjord,* was in port with two cousins of mine—Erma and Mabel—on board. I had dinner with them on the ship and returned to duty two days later.

On 10 March Lyle and I took a gunship to Phu Bai and then to the A Shau Special Forces camp. We were shot at twice but took no hits. The camp was surrounded by Viet Cong but the airstrip was operational. In my estimation, the H-34s at Phu Bai weren't doing a lot to get in there. The weather wasn't the best, and there was a mountain range between Phu Bai and A Shau, but nevertheless I became impatient and got in my Huey gunship and flew down there to talk to the squadron. After conferring with the CO I said, "Well, I'll check the weather." I climbed up through it and was on top at 8,000 feet. Lyle recalls:

"It was apparent the general had been going out to Special Forces camps for quite some time. Invariably, the captain commanding would salute and greet him by name and I'd say to myself, 'He's been here before.' One time we landed and the guy said, 'General, we really want to thank you for the generator.' Usually Marine Corps generals don't give generators to the Army. I think General Carl also got a refrigerator for some outpost.

"To tell the truth, I never liked the Army, nor did I have any respect or admiration for it. But when I see a Green Beret in a bar, I buy him a drink. There was no way I ever wanted to spend a night in one of those places they held."

All of a sudden there was an Air Force bird-dog alongside and we got on a common frequency. He said, "We have a pilot down at A Shau. Would you go in and pick him up?" An A-1E Skyraider, call sign Surf 41, had crashed on the strip, which was being overrun by Viet Cong. There was no time to waste.

I said I would give it a go if I could get down through the overcast. He said there was a hole near A Lui about twenty miles from A Shau, so I followed him down. I lost him descending through the hole and was only about five minutes out of A Shau when he said another A-1 had just picked up the downed Skyraider pilot. Maj. Bernie Fisher of the 607th Air Commandos landed, avoided debris and small arms fire, pulled his wingman—a fellow named Meyers—into the cockpit, and took off through more gunfire. It was a Medal of Honor performance.

I went across A Shau just below the overcast, about 1,200 feet; the Viet Cong opened up but I collected no hits. When I returned to Danang there was a big conference with several ranking officers—I was probably the junior one present. Gen. Lew Walt asked each of us what we thought about the situation. I said I thought we should go in with an effort to get the Special Forces and friendly Vietnamese out of there, rather than abandon them. I said, "General, we'll lose one out of four helicopters but we should try."

Lew agreed. But first he asked me about the weather and I told him. He pressed the matter and asked, "How do you know so much about the weather there?" I replied, "Well, I admit I was in the area in a Huey." He didn't like that too well—brigadiers were supposed to do other things—

though he didn't say much about it. But I believe he would have done the same thing under similar circumstances.

We sent in twelve H-34s and I think we lost three, including the one Lieutenant Colonel House, the squadron CO, was flying. But he and his crew made it back; I don't think we lost anyone. Unfortunately, the Vietnamese panicked and mobbed the H-34s, which had to open fire on some of them so the helos could get airborne. Even so, I think they retrieved nearly seventy people.

On 11 March I got a look at the first CH-46s we received in Vietnam, and I took a familiarization hop the next day. They were the newest airlift helos, and we badly needed them to replace the H-34s. War correspondent Michael Herr wrote, "The -34 had a lot of heart," but despite the lowest loss rate of any helo in-country, it was past its prime. It couldn't lift as much as we liked, which meant that more sorties were necessary to deliver a given number of troops or supplies.

In mid-month I flew a close air support mission in an F-8E, dropped four 260-pound frags, and did some strafing with the 20-mm guns. That evening Ann Margret and her party were our guests for dinner. She sat between Keith McCutcheon and me at the head table, and I don't think either of us said five words to her. We didn't know what to talk about! Maybe it was just Vietnam. That morning I had been bombing and strafing in a supersonic fighter, and that evening there I was dining with a glamorous, gorgeous redhead. She must have wondered what happened to young marines who grew old.

The next day Lew Walt walked up to me while we were both visiting a combat zone. He pulled a message from the Commandant from his pocket. In essence, it said that general officers were prohibited from participating in any type of combat operations. Lew said, "This applies to you in particular, Marion." I said, "Yes, sir. However, General, I believe it applies to you, also." I knew that he had been up to the front lines many times during his tour.

It was an order I more or less obeyed. I had flown about forty combat missions in Hueys and H-34s, as well as A-4s, F-4s, and F-8s, but from then on my emphasis was administrative. I was tagged by General Walt as senior member of two boards of investigation. One dealt with the Special Forces at A Shau and the other concerned Lieutenant Colonel House's squadron. The proceedings lasted two days, with Generals Westmoreland and Krulak in attendance. As far as I know, no action was taken for or against either party. I suspect some medals were awarded long after I had returned to the States, but by then my last war was behind me.

CHAPTER TWELVE

The Sheriff of Cherry Point

It took me a while to get out of Vietnam. During late March and early April I cleaned out my desk, make some visits, laid some bricks for a local school, and even gave a speech to MAG-12 at Chu Lai. I would have much rather bombed Hanoi in an F-4 or hunted MiGs in an F-8, but even generals have to take orders.

Lyle Prouse left with most of my gear on 6 April and I followed three days later. My trek homeward was made via C-130 and Pan Am 707, and Edna welcomed me home to Honolulu. It was a busy time in Hawaii, and I didn't eat dinner at home for a week. However, it was nice to see Dick and Mona Tregaskis again. Dick was still writing, though we agreed that Vietnam didn't provide him with the type of stories he found on Guadalcanal because he couldn't go where he pleased in Vietnam.

In May I settled in at MCAS Cherry Point, North Carolina, where I relieved Norm Anderson as base commander. My chief of staff was Col. Bob Smith. There were other friends in the area, too, including Bill Kellum, my former exec in MAG-33, who was living in New Bern. That was one nice thing about Marine Corps aviation—it was still small enough that you found people you knew wherever you went. The service had a genuine sense of family about it.

Edna took some extra time before moving from Hawaii, so I bached it for a while. For lack of anything else to do, I started patrolling the base from 2300 to 0200 every night. It seemed a good way to learn firsthand what was going on, and I wasn't happy with what I saw. I didn't think the MPs were doing a very good job, so I got on their backs and continued to do my patrolling.

Normally I drove an official car by myself. I would stop an offender personally, and then call the MPs and have them handle the situation. On my first "arrest" I had the help of a staff driver. We pulled over a sergeant who was driving while under the influence of something, weaving back and forth; when the MPs opened his trunk they found a bunch of stolen rifles.

I figured nobody that stupid had any business being a Marine Corps non-commissioned officer.

The episode with the stolen rifles was typical in some ways. There was no earthly means to ensure compliance with weapons regulations—military or civilian. For instance, Cherry Point required all personally owned firearms to be registered, as well as knives of certain blade lengths. But during my entire tenure as base commander, not one weapon involved in a crime had been properly registered. Obviously, if "gun control" won't work in the Marine Corps, it won't work in civilian life, either. I never did learn the intended purpose of registering personal weapons on base. As far as I could see, the only reason would have been eventual confiscation.

That first incident only encouraged me to continue my patrols, and later I learned I had a reputation as "the sheriff of Cherry Point." Other events forced me further afield. There was an offbase nightclub called Dino's that stayed open after the on-base clubs closed. I stopped in following the official Marine Corps birthday festivities 10 November, and saw a lot of young marines obviously under the influence. We had experienced numerous drunk-driving incidents between Dino's and the base, and sometimes even on base. On one occasion I paced a marine doing ninety in a forty-five zone. So I declared Dino's off limits.

The club owner was an ex-Marine, and he sued me for about a half-million dollars—something about "depriving him of a livelihood." Because I had acted in an official capacity, I was not liable and the case was thrown out. Eventually the place went out of business.

By mid-June I had been baching for a month and was having difficulty getting Edna interested in leaving Hawaii. She had a nice place on the beach—still in generals' quarters—and a lot of friends. Whenever I brought up the subject of moving she offered some excuse for why she couldn't leave just then. So I called Brute Krulak and sought his help. I said I needed Edna at Cherry Point, where she belonged. General Krulak just laughed and asked what he could do, especially since he and Edna were great friends. I said, "Well, General, you can dispossess her."

Edna arrived on 6 July.

Four years at Cherry Point allowed us an unusually stable family life. Lyanne graduated from high school at New Bern in 1967, while Bruce attended Fort Union Military School for two years and then returned to regular high school, graduating in 1970. I had time to develop my archery skills—I usually hunted with Col. "Chief" Leu. A civilian named Carl Bynum became my fishing partner, and we also hunted doves and ducks together. Unfortunately, my friendship with Carl ended when he was sent to prison for shooting and wounding a man. I learned his other pals were under close scrutiny by local game wardens for poaching, so it seemed prudent to conclude my association with that bunch. I learned to pick my associates more carefully!

Cherry Point had some outlying fields, so I did a bit of flying to remain proficient. Before Christmas 1966 I went to Marine Corps Headquarters, landing at Andrews AFB in a US-2B. The ceiling was about two hundred feet with half a mile visibility. With snow. Below freezing. I wanted to proceed to Quonset Point, Rhode Island, and told my copilot, a major, when I would return from Headquarters. I asked him to have the plane deiced. When I got back, he said it was ready to go.

As soon as the wheels came off the ground the nose pitched up and I ran full forward elevator trim. But before I knew it the nose was well up and we stalled at about 1,100 feet. As soon as the nose dropped off, I applied left aileron and left rudder, forcing a wingover. By then I was getting smart real quick. I dropped the landing gear, leveling at 400 feet, still in the soup. Meanwhile I told the copilot to get me on ground-controlled approach. He wanted to declare an emergency and began calling Mayday, but I told him to forget it. About that time we went past the water tower, just missing it in the gloom. That set off the major again, calling Mayday some more.

We got on the ground-controlled approach downwind leg and I said we had to make an emergency landing. They brought me right around with a short pattern and I plunked the Trader down on the runway. While taxiing, the tower asked if I wanted my flight plan canceled, and I said no. I thought I knew the problem, and expected I'd be back in a half-hour.

I turned to the copilot and asked if he had visually inspected the top of the horizontal stabilizer for ice. He said no. The horizontal stabilizer was about eight feet high, and you couldn't see it from the ground. We shut down and checked the tail, and sure enough, we had almost an inch of snow and ice over the stabilizer and elevator. We called the deicing truck, which applied a spray job, and we took off without trouble.

That was the first and last time the major flew with me. For some reason he never made lieutenant colonel.

Next time I got a new copilot.

I got my second star in August 1967, when Gen. Leonard Chapman "frocked" me. Although my billet was a major general's and I wore the stars, I did not get equivalent pay for several months. The same thing had happened when I made brigadier general.

A year later I took over the 2nd Marine Aircraft Wing, and Keith McCutcheon came down from Washington, D.C., for the ceremony. Three days later I began refamiliarization in tactical aircraft, beginning with the F-4J. The Phantom remained my aircraft of choice, especially because of its cross-country capability. I could make it from Cherry Point to El Toro in four and a half hours, including a thirty-minute fuel stop at Tinker AFB, Oklahoma. It sure beat twelve hours in the S-2!

In January 1969 I represented the Commandant at ceremonies honoring two Medal of Honor winners. One was Maj. Steve Pless, one of those rare individuals without fear in his makeup. He had rescued some marines from behind enemy lines when nobody else was even willing to try. The other was an Air Force pilot, Lt. Col. Bernie Fisher, who had gone into A Shau to rescue another A-1 pilot back in 1965. He had beaten me there by about five minutes.

The sidelight to this story was our arrival at Newman, Georgia, for the ceremony. I had authority to take Edna in my C-54 and the 2nd Wing's band in another C-54. The Air Force base at Marietta said the weather was ceiling zero and one-sixteenth mile visibility, and asked my intentions. I said I intended to land.

My copilot was a master sergeant named Todd, with most of his 10,000 hours in transports. He was nervous but I was flying the airplane. The only reason the tower let me try to land was that their records showed there was a Marine Corps flag officer aboard. If it had been an Air Force aircraft they never would have let me make the approach, but since it was Navy they didn't care.

On final I told Todd, "Let me know when you pick up the lights," since my attention was riveted on the instruments. Todd said he had the lights and I asked, "One row or two?" He replied, "Only one," so I took a waveoff and the tower asked what I was going to do. I told them I'd come around again.

This time Todd called the lights and said, "I have them both." I completed the landing and rollout, but had to wait for the follow-me jeep because I couldn't see the taxiway. They wouldn't allow the other C-54 to land, so the band returned to Cherry Point.

There's a postscript to the story. Not long after the award ceremony, Steve Pless was going home from a Pensacola beach party, apparently driving his motorcycle like he flew a helicopter. There was a drawbridge on his way home that started rising while he approached at about 100 mph. Evidently he didn't think he had time to stop, so he kept going. The barrier was coming down but he squeaked under it and went up the bridge, intending to jump the river and land on the other span. He almost made it. But he hit the abutment and was killed instantly. Oddly, I knew an insurance agent who had sold Pless a $40,000 life insurance policy just forty days before. The agent said if he had known what kind of guy Steve was, he'd never have written the policy. But it paid off in full.

The 2nd Marine Aircraft Wing flew a wide variety of types, and eventually I checked out in all of them, including A-6 Intruders, OV-10 Broncos, and CH-53 Super Stallions. I also got a few flights in "Brand X" types, including the Air Force's F-111. Cross-country opportunities included events such as the annual Tailhook symposium and the American Fighter Aces Association. I'm told I was the last active-duty ace to fly a combat aircraft to the latter event when I took an F-4 to Colorado Springs in 1969 and accepted the group's presidency.

That year I got to fly something even more exotic. During a trip to Nellis AFB I was flown in a T-33 to a secret desert site in order to fly a MiG-17 that had been captured in the Middle East. I had two hops for 1.7 hours, neither of which was permitted to be logged. On the first flight I tangled with a couple of F-8s, and on the next it was A-4s. I was really impressed with the MiG-17's maneuverability. You could stall going straight up and let it fall off one way or the other, but within 1,500 to 2,000 feet you had full control. The only difficulty was the cockpit size, as pilots over six feet were pretty cramped.

I asked about maintenance problems and was told there just weren't any: "Would you believe we've flown 100 flights with this airplane, and the down time has been less than an hour?" That included some flights firing the 23- and 37-mm cannon.

About the middle of the day the staff began pulling the MiGs off the line into nearby hangars. I asked about the procedure and was told that a Soviet reconnaissance satellite was due overhead. Because it was a clear day they didn't want any pictures to be taken of the captured MiGs.

Eventually the "secret" MiGs became a source of embarrassment. Everybody in military aviation knew they were there, and *Aviation Week* ("Aviation Leak") magazine even published some photos, but the Air Force continued to deny their existence.

In the spring of 1970 I took an F-4J to the McDonnell factory in St. Louis and parked on the company ramp. There was a flag officer conference with a dinner that night, hosted by Sandy McDonnell. As I traveled alone, I had nobody in the back seat, and this caught the attention of a couple of Air Force generals. They asked, "How do you get away with it?" "Get away with what?" I replied. "Do you have authority to fly all by yourself?" they asked incredulously. My response was, "Sure, why not?"

The Air Force officers said they had to have a safety pilot. The Marine Corps policy impressed them. They didn't understand how we could get away with something they couldn't, but as far as I know the situation remains unchanged. At least one Marine Corps aviator general has flown F/A-18s against Air Force pilots, listing himself as an overage captain in order to avoid the hassle of courtesy calls from other generals!

In mid March 1970 I learned I would be losing the wing after my two-year stint. I knew I was going to Marine Corps Headquarters, and shortly I learned that I would become the Inspector General of the Marine Corps. It was hardly an ideal place for an aviator. I gave some thought to retiring and discussed it with Edna and with General Walt. However, I told him I would stay. I thought I had a reasonably good chance of getting a third star, and if I didn't get it under General Chapman I might succeed under the next Commandant. I hoped that Lt. Gen. W. K. Jones, whom I considered a personal friend, would be the next Commandant.

We prepared to return to our house near Quantico. I would have a real nice corner office in the Donato Building in Rosslyn, near the Key Bridge in Washington, D.C. My staff and I would be off by ourselves, not subjected to many of the visitors to Headquarters.

On 3 May I had a call from Maj. Gen. E. E. Anderson in Headquarters. He was also an aviator, somebody I had known for quite a while, and obviously going places in the Marine Corps. Not much later he was a lieutenant general, and I think he went from two stars to four in a year. At that time I didn't realize that I would be working for him later.

In August I made the first of numerous inspection trips, a quick look at El Toro. All the legwork was done by three preliminary teams, each headed by a colonel. That same month I was off to Iceland in a C-54, inspecting the Marine Corps barracks at Keflavik. The rear admiral there was fellow ace Mike Hadden, who arranged a visit in a charming young couple's home. Next day I proceeded to Northolt, U.K., and the following morning I was jogging in Hyde Park. It was one of the best places I have ever run because of the open area and good turf. Next stops were in Scotland, Italy, and Spain. I really got around—even to the point of inspecting the Marine Corps guard compound at Camp David.

In February 1971 Edna and I attended an affair at the Patuxent River officers' club, where several hundred people were invited to celebrate "Gorgeous George" Watkins's 10,000th flight hour. The walls were covered with life-size photos of George throughout his career. At the time I had more than 13,000 hours, and George's mission in life was to finish his military career with more flight time than I had. Although he gave it a good try, he never caught me. At Tailhook '85 he told Barrett Tillman that he was still trying "to catch that damn Marion Carl!" I admit to a chuckle over that one.

One of the pleasant benefits of my otherwise dreary job was meeting interesting people. And certainly one of the most interesting was a former enemy. During an inspection trip in the Far East in May 1971 I stopped in Tokyo and registered at the Sanno Hotel. There I was met by Gene Valencia, a twenty-three-victory naval aviator and cofounder of the American Fighter Aces Association. Gene and a writer named Bailey had arranged for me to meet with eight former Zero fighter pilots. The senior member of the group was Saburo Sakai, Japan's top living ace.

Saburo spoke no English, but one of the other pilots spoke it well. We all went to a restaurant a couple of blocks from the hotel and had a fine dinner with a lot of sake. I was at the head of the table, and Sakai's former wingman was at my side, so I could talk to him and learn more about Saburo. I asked what it was like to fly with Sakai. "Oh, he was a very good flier," said the wingman, who is a big fellow while Saburo is a little guy. The wingman said, "Saburo was a tough leader. If he didn't like the way I had flown, when I climbed out of my airplane he came up to me and hit me as hard as he could. I didn't dare hit back."

I thoroughly enjoyed that evening. Like most pilots, I bore no grudge against former enemies and even discovered we had a lot in common. I got to see Saburo again many years later at the Yakima, Washington, air show in 1986. Most recently we appeared together at the Admiral Nimitz Center in Texas, recalling old times at Guadalcanal during the Admiral Nimitz Center history symposium in 1990.

In May 1971 I returned from the Far East through Hawaii and stayed with Bill Jones, then commanding Fleet Marine Forces Pacific. Bill and I always got along well. If he had made Commandant—as I think he should have—things would have turned out far better for the Marine Corps. In my opinion, the only reason Bob Cushman became Commandant was that he had been Nixon's naval aide. Although Bob was a fine person, he had been out of the Marine Corps for two or three years as deputy CIA director, and when he came back in it was too late for him to catch up on things. Sadly, it was another result of the Nixon administration's partiality to political favorites.

The one official I did learn to respect was Secretary of Defense Melvin Laird. He was a hard worker and seemed willing to listen to professional opinions, even if he was uncertain of them. It was a pleasant change from the Johnson-McNamara regime.

A Navy officer I appreciated was Adm. Tom Moorer—an astute individual who deserved his position as chairman of the Joint Chiefs of Staff. I always felt that one of the marks of an officer was how he treated subordinates, and Admiral Moorer always recognized me and stopped to talk.

In 1972 Marine Corps politics began getting in the way of efficiency—at least that's how I saw it. Bob Cushman pulled up E. E. Anderson—formerly his chief of staff with the 3rd MAF in Vietnam—to assistant commandant. There was no doubt that Anderson was capable, and it appeared to me that he was running the Marine Corps almost singlehandedly owing to Cushman's unfamiliarity with things.

I was interested in remaining on active duty as long as I was in contention for a third star. I didn't want to drift along like deadwood, just filling a slot without the chance to contribute something. Therefore, I was pleased when Lt. Gen. Lou Wilson asked if I would be willing to become his deputy with the Fleet Marine Force on Okinawa. At one time I had been senior to Lou but now he was senior to me, and it didn't bother me in the least. I told him I would be pleased to work for him. His reputation as a combat marine was second to none (he had won the Medal of Honor as a twenty-four-year-old captain), and he got on well with Cushman.

On the evening of 27 November I got a call at home from Maj. Gen. Carl Hoffman, director of personnel. Orders had been published for me to go to Korea as senior member of the armistice commission. It took me about five seconds to realize that my career was over. I only had a couple more years before retirement, and the year in Korea would take me out of the running for promotion. It was unusual for a general to have his orders published before being verbally informed of them, but I didn't say anything to Carl because it wasn't his doing. I knew that I would brief Cushman about my latest trip in a few days, so I decided to wait.

On 1 December I briefed the Commandant on a recent inspection and added that I intended to retire. This visibly upset him because the orders had been published. I said, "I have twice in the past six months told General Anderson that if I were not being considered for a third star, I wanted to retire. The last time was only a couple weeks ago." Cushman replied, "Don't worry. We'll take care of it." The upshot was that I did in fact retire early, but, fortunately for the Marine Corps, Lou Wilson got his fourth star and became the next Commandant.

In May I had my last fling in uniform, attending the Pax River test pilot reunion with Butch Davenport. Then came the difficulties of actually getting out of the Marine Corps. Headquarters asked if I wanted a retirement parade at "Eighth and Eye," the traditional Marine Corps parade ground in Washington, D.C. I asked them to forget it—at that point I was fed up with the way the Marine Corps was handling things. My retirement took place in the Commandant's office on 31 May 1973, and so ended my thirty-five years in Marine Corps aviation.

CHAPTER THIRTEEN

Retirement and Hunting

In the years after I retired I tried my hand at a number of endeavors, including a wood-products business in LaPine, Oregon, during 1978 and a brief fling with a charter aircraft outfit at Spokane, Washington, in 1979. I kept my trailer limbered up, crossing the country from Quantico to Oregon and back on business, hunting, and pleasure trips.

Before we moved from Virginia to Oregon in 1980 I took one last shot at Quantico's turkey hunting. Most of my gear was out West so I wore blue jeans and a khaki shirt and lugged a shotgun. I had been out for a while without seeing anything, so I headed back to the car a half-mile away. En route I passed an abandoned farmhouse with some daffodils around it. They were in bloom, really attractive, so I thought, "Well, I didn't get a turkey but I'll get myself some flowers for Edna." I picked a whole handful—a real nice bouquet—and was walking up a dirt road a couple hundred yards from the car when a tom walked directly in front of me. He was on a slight curve near the road and didn't see me. I trotted up the road to where he had disappeared into trees on the other side and waited. Pretty soon he waddled into the open about twenty yards away. I couldn't believe it. I didn't even drop the flowers—just pulled down on him and fired.

When I dragged the bird up to the car there were two other hunters in full turkey-hunting uniform with camouflage and face paint. They hadn't shot anything. But when they saw me dragging the tom, trying to hold onto my gun and the flowers, naturally they became curious. They asked if I had just fired and I said yes. Then they asked about the daffodils and I replied, "They're for my wife. She'll probably appreciate them more than the turkey." (Which she did.)

The idea for our move to Oregon had been sparked by my friend Joe Rees, who had settled on the North Bank of the Umpqua River near Roseburg, Oregon. As a former test pilot, Joe was analytical in all things, and after he retired from Grumman he conducted a thorough evaluation of possible retirement spots. He calculated that Roseburg fit him for climate, hunting, and fishing, and Edna and I liked what we saw there, too.

Joe and I began a series of annual elk hunts in the Blue Mountains of northeastern Oregon at a site recommended by the Jack Tillman family of Athena. I usually filled my tag, but the occasion I remember best had nothing to do with bagging game. It was an exercise in heliborne logistics.

During the 1987 expedition a couple of the Tillmans' friends decided on a deluxe affair. Both these fellows—Johnny and Morrie—have a flair about them, and in collaboration with John and Andy Tillman they established an after-hunt affair that I swear the Safari Club couldn't surpass. Their father, Jack, didn't even know about it, which was the whole idea. The event was in honor of his pioneering Table Springs as an elk camp back in 1947.

Joe, Butch Davenport, John Wheatley, Bill Meyring, and I tramped back into camp at dusk and noticed a neon sign glaring through the darkness. Though we had seen a helicopter fly over that afternoon, we didn't think much of it. But the neon seemed awfully unusual, so we investigated. It was a Miller Beer sign, suspended from a pine tree and run by a portable generator. A table and bar had been established, with Johnny and Morrie serving as waiters. In fact, Morrie was decked out in a tuxedo (complemented by red tennis shoes), complete with white towel draped over one arm. They were pouring aperitifs, and drew an immediate crowd. A hunter from another camp saw the setup, smacked his lips in anticipation, and inquired when the bar would begin serving. "We're not open yet," snapped Johnny, and the interloper sadly trailed off.

The dinner was just superb, featuring shrimp cocktail, onion soup, Salmon Alaska with salad, and a variety of liqueurs and desserts. To do justice to the menu would require more space than is allowed here.

We were all astonished by the magnitude of the undertaking, but none more than Bill Meyring. Bill was a superior chef himself, but he didn't mind being displaced for one meal as the camp cook. Somehow Johnny and Morrie had crammed the entire meal, with supporting gear, into a little Bell 47 and made a successful "vertical insertion" into the nearby "landing zone." I learned that Morrie had flown Army helos in Vietnam, so it wasn't entirely surprising.

We later learned that the duty hostess had canceled at the last minute. Morrie's girlfriend, MiSchelle, a private investigator, had appeared in *Playboy* not long before, and had intended to serve as waitress—complete with bunny outfit. That fall was unseasonably warm in Oregon, without snow, so she could have worn the skimpy costume without problems.

The story had a bittersweet ending. Butch made a videotape of the whole campaign and sent copies to all the principals. But it was his last hunt; he finally lost a long battle with cancer in September 1989. I've never had a better friend—and neither has anyone else.

There's a saying that getting older beats the alternative, and in some ways I'm living proof. I missed elk season in 1981 after falling off the roof of our new house, and again in '88 when I had hip surgery. The cartilage had worn away until it was bone on bone, and I was laid up for a few months. But now I'm back to climbing hills and taking long walks, so I figure as long as there are replacement parts I may last forever!

From my desk at home I can see most of the activity on the river, and I never tire of watching the wildlife. I've seen a bald eagle try to catch one of my tame ducks, and it's quite a process.

When I see what's happened to the earth, I understand how Charles Lindbergh must have felt. After growing up in the Pacific Northwest and living here again for the past twenty years, I've been sickened by the way we've treated the earth. The timber industry, so important to the region and to my native Oregon, has been allowed to do terrible damage for reasons of economic and political expediency. Clearcutting in some areas deprives the earth of roots that hold the soil in place, and in any heavy rain I see muddy waters taking irreplaceable topsoil downstream. Air pollution now fouls areas where only light haze existed in my cross-country flights of the '40s or even '70s.

Since my retirement in 1973 I have often been asked what I think about current events in the service or in military aviation. Usually I insist that I am too far removed from that world to provide a meaningful assessment. However, I don't mind voicing some opinions based on my experience.

One of my biggest pet peeves is the way we handle awards and decorations. Based on observations from three wars, there is no doubt that combat medals are dispensed in much too random a fashion. For instance, in Vietnam I made 12 flights in fighter-bombers and 114 in helicopters, which would have qualified me for about twenty-five Air Medals. But I neither requested nor was awarded such decorations because I felt they were meaningless—especially because almost any helo flight in-country could be lumped into the combat category because you were liable to draw sporadic small-arms fire just in transit.

The problem dated at least from World War II. Among Navy and Marine Corps fighter pilots of 1942–43, Medals of Honor often were presented for downing five or more enemy aircraft in one mission. But by mid to late 1944, the feat was so routine that a Navy Cross or Silver Star became almost automatic. This medal inflation only confused matters and, to an extent, denigrated the higher decorations awarded earlier on.

Consequently, I think that especially today—when wars are likely to be intense but short—we should wait until all the results are in, and then, after proper assessment, present awards. After all, in the profession of arms the only criterion that matters is damage to the enemy. All else—including life-saving and outstanding support of combat troops—should be recognized relative to that.

There appears little doubt that women are going to be assigned to combat aircraft in the 1990s. I am convinced that it will prove a costly mistake, though not for the usual reasons attributed to male chauvinism. There is no question that many women can fly military aircraft extremely well, and over a period of decades perhaps male fliers really will learn to accept women within the special atmosphere of a tactical organization.

However, thousands of years of human development will not be changed anytime soon. Men will continue to be either protective of or resentful toward women in positions of physical danger. And while sometimes I hear that history proves the case for women in combat, look closely. The Soviets and Israelis both employed women in combat during an acute manpower shortage, and both nations immediately ceased their programs when the crises had passed. But we have no such shortfall; in fact, we are well into a period of drastic reduction in the armed forces.

The problem, as I see it, is political. Interviews show that very few women now in the armed services have any desire for combat. Those who do seek hazardous duty should be as well qualified as their male counterparts. But here is my prediction: in order to appear politically correct the Defense Department will be directed by Congress to accommodate as many women as possible who want to qualify for combat duty. If we begin losing people who are physically or otherwise unsuited for war, the fault will lie with political opportunists combined with weak service chiefs unwilling to fight city hall.

The "Tailhook scandal" of 1991–93 is a case in point. As a longtime flag officer participant, I know the value of the Tailhook Association, its annual symposium, and its professional journal. However, following Tailhook '91 the Navy's civilian and uniformed leaders unnecessarily botched what should have been a straightforward investigation of allegedly abusive behavior by a few individuals. Instead, under political pressure that nobody in the Navy had the character to withstand, a worthwhile organization was torn apart and aviator morale nosedived to probably the lowest level since World War II.

For those still seeking the peace dividend, I think we need to keep our perspective. It is far better to invest in realistic training and reliable, proven equipment than to pay off GI insurance policies. As long as there are two humans on this planet, they will find something to fight about. That's not merely a justification of the way I spent thirty-five years of my life—it's simply a fact. Unfortunately, one of the greatest reasons for continuing warfare is religion, and the end of killing in God's name is nowhere in sight.

In looking back, I realize how fortunate I was to enter the Marine Corps when I did. I was fortunate in that I was born on a dairy farm where hard work was expected six and a half days each week, with no vacations. My father died when I was in high school, yet my younger brother and sister and I graduated from college. I went to naval flight training and a Marine Corps career, which provided me with opportunities that I never expected or even dreamed of.

I was blessed with the physical and emotional characteristics needed to make a career of the one endeavor I wanted to pursue on this earth—flying fighter and experimental aircraft. When I recall that I began in a 40-horsepower Piper Cub and finished in an F-4 Phantom with an equivalent of 60,000 horsepower, I have nothing to regret.

It is true that my life was on the line many times, but I never lost any sleep over the close calls. Of the seven reportable accidents, four of the aircraft never flew again and two were never seen again. While in Flight Test, four pilots who flew my project airplanes were killed. In addition, many of my squadronmates were killed or wounded in combat, yet I survived to fill eight logbooks and eleven diaries.

My two main wars—the Southwest Pacific and South Vietnam—allowed a degree of personal involvement that just is not possible today. Between 1942 and 1965, in the aircraft I flew, pilots had the personal satisfaction of actually seeing their results at close range. And while the ultimate outcome in Southeast Asia was frustrating, maddening, and perhaps even avoidable, I recognize that nowhere else could a general officer participate to the extent that I did.

I still like to think that I helped make a difference, avoiding unnecessary losses that characterized so much of that ill-directed war—and even managed a fair amount of shooting as well.

After all, that's what the United States Marine Corps trained me to do.

Bibliography

Fails, William R. *Marines and Helicopters 1962–1973.* Washington, D.C.: U.S. Marine Corps, 1978.

Frank, Richard B. *Guadalcanal: The Definitive Account of the Landmark Battle.* New York: Random House, 1990.

Lundstrom, John B. *The First Team.* Annapolis: Naval Institute Press, 1984.

Olynyk, Frank. *USMC Credits for Destruction of Enemy Aircraft in Aerial Combat, WW II.* Privately printed, 1981.

Rawlins, Eugene W. *Marines and Helicopters 1946–1962.* Washington, D.C.: U.S. Marine Corps, 1976.

Sherrod, Robert. *History of Marine Corps Aviation in World War II.* Washington, D.C.: Combat Forces Press, 1952.

Tillman, Barrett. *Corsair: The F4U in World War II and Korea.* Annapolis: Naval Institute Press, 1979.

———. *Wildcat: The F4F in WW II.* 2d ed. Annapolis: Naval Institute Press, 1990.

Tregaskis, Richard. *Guadalcanal Diary.* New York: Random House, 1943.

Thanks to the following individuals who contributed recollections or documentation: Art Adams, Joe Foss, Bill Hardy, John Lundstrom, Lyle Prouse, and Mark Gatlin at Naval Institute Press.

Index

NOTE: Military personnel for whom no armed service is indicated belong to the U.S. Marine Corps or the U.S. Navy.
Ranks shown are those held by individuals at the first mention of them, and not necessarily the highest achieved.
The abbreviation *MC* stands for Marion Carl.

About the Authors

Marion Carl's career in Marine Corps aviation spanned thirty-five years, from 1938 to 1973. He retired as a major general with 14,000 flight hours in every type of aircraft: from biplanes, seaplanes, and helicopters to jets and rocket-powered experimental models. Besides flying combat missions in World War II and Vietnam, he has held both the world's altitude and speed records. Credited with eighteen and one-half aerial victories, General Carl is the Marine Corps's seventh-ranking fighter ace. He lives with Edna, his wife of fifty years, near Roseburg, Oregon.

Also a native Oregonian, Barrett Tillman learned to fly as a teenager. He has written seven other Naval Institute Press books in addition to three published novels, and has won three aviation and naval history writing awards. As a professional author he divides his time between Athena, Oregon, and Mesa, Arizona. He is active as a contributing editor to *The Hook* magazine and serves as a trustee to the Association of Naval Aviation.